James Edward Henry Gordon

School Electricity

James Edward Henry Gordon

School Electricity

ISBN/EAN: 9783744791113

Printed in Europe, USA, Canada, Australia, Japan

Cover: Foto ©berggeist007 / pixelio.de

More available books at **www.hansebooks.com**

SCHOOL ELECTRICITY.

WORKS BY THE SAME AUTHOR.

Now ready, Second Edition, 2 vols. 8vo, 700 pages, with 72 full-page Plates and 313 Illustrations in Text. Price £2 2s.

A PHYSICAL TREATISE ON ELECTRICITY AND MAGNETISM. This Edition has been Rearranged, Revised, and greatly Enlarged, both in Text and Plates, to bring it up to the present date.

"A book which every one emulous of the name of an Electrician should possess."—*The Electrician.*

Demy 8vo, with 24 full-page Plates and numerous Illustrations in the Text. Price 18s.

A PRACTICAL TREATISE ON ELECTRIC LIGHTING.

"Without doubt it is the most valuable work on Electric Lighting that has yet appeared."—*The Electrician.*

With Illustrations, 16mo, limp, 3s. 6d.

FOUR LECTURES on STATIC ELECTRIC INDUCTION, DELIVERED AT THE ROYAL INSTITUTION.

Crown 8vo, cloth, 2s. 6d.; without Answers, 2s.; Answers separately, 6d.

NUMERICAL EXERCISES in CHEMISTRY (Inorganic) 650 Examples. By T. HANDS, M.A., F.R.A.S., Science Master in Carlisle Grammar School, and late Scholar of Queen's College, Oxford.

Crown 8vo, 330 pp., with very numerous Illustrations, half roan, cloth sides, 5s. post free. Second Edition.

A SCHOOL COURSE ON HEAT. By W. LARDEN, M.A., Assistant Master in Cheltenham College; late Science Scholar, Merton College, Oxford. Adopted at Rugby, Clifton, Bedford, Birmingham, and many other Public Schools.

Crown 8vo, cloth, very numerous Diagrams, &c., 2s.

A PRIMER OF ORTHOGRAPHIC PROJECTION; or, Elementary Practical Solid Geometry Clearly Explained, with numerous Problems and Exercises. Specially adapted for Science and Art Classes, and for the use of Students who have not the aid of a Teacher. By Major G. T. PLUNKETT, Royal Engineers.

Crown 8vo, with numerous Diagrams, 320 pp., cloth, 7s. 6d.

A TREATISE ON NAUTICAL ASTRONOMY. By JOHN MERRIFIELD, LL.D., F.R.A.S., F.R. Met. Soc.; Head Master of the Navigation School, Plymouth; Author of "Navigation and Nautical Astronomy," "Magnetism and the Deviation of the Compass," &c., &c.

London: **SAMPSON LOW, MARSTON, SEARLE, & RIVINGTON,**
188, Fleet Street, E.C.

BY

J. E. H. GORDON, B.A. Camb.

MEMBER OF THE INTERNATIONAL CONGRESS OF ELECTRICIANS, PARIS, 1881;
MANAGER OF THE ELECTRIC LIGHT DEPARTMENT OF THE TELEGRAPH
CONSTRUCTION AND MAINTENANCE COMPANY.

"I have long held an opinion, almost amounting to conviction, in common, I believe, with many other lovers of natural knowledge, that the various forms under which the forces of matter are made manifest have one common origin; or, in other words, are so directly related and mutually dependent, that they are convertible, as it were, one into another, and possess equivalents of power in their action."—FARADAY, *Exp. Researches* (§ 2146), 1845.

London:

SAMPSON LOW, MARSTON, SEARLE, & RIVINGTON,

CROWN BUILDINGS, 188, FLEET STREET.

1886.

[*All rights reserved.*]

PREFACE.

In the course of the last twenty years the Science of Electricity has emerged from the theoretical to the practical state. Special branches of it have been developed, while the importance of other branches has diminished. The object of the present work is to give schoolboys a knowledge of electricity which, however incomplete, shall be of a useful kind.

For this purpose the work commences with the study of electric currents and Ohm's law, and the teaching is of the same kind as that which I have been in the habit of giving to members of my staff, and to men employed in electric light work. At the same time I have given such details of theory as will, I hope, show students that a science of electricity exists, and stimulate them to study it. I have further endeavoured to show the intimate connection which exists between electricity and light, heat, and mechanical energy.

CONTENTS.

CHAPTER I.
ELECTRIC CURRENTS.

	PAGE
Electric currents	1
Questions	3

CHAPTER II.
LAWS GOVERNING THE FLOW OF ELECTRIC CURRENTS.

Ohm's law	4
Speed of electric currents	5
Questions	6

CHAPTER III.
MAGNETS AND ELECTRO-MAGNETS—LINES OF FORCE.

Magnets	7
Action between **magnets and currents**	8
Electro-magnets	10
Telegraphy	11
Lines of force	12
Yokes and magnetic **circuits**	14
Questions	16

CHAPTER IV.
RELATION OF ELECTRIC CURRENTS AND ELECTRIC PRESSURE TO MECHANICAL FORCES.

Illustration from water-wheel	17
Foot-pounds	18

vi *Contents.*

	PAGE
Horse-power	18
Electrical horse-power	19
Questions	20

CHAPTER V.

ELECTRICAL UNITS AND THEIR RELATION TO EACH OTHER, AND TO THE HEAT AND WORK UNITS.

Current.—Ampère	21
Pressure.—Volt	22
Resistance.—Ohm	22
Quantity.—Coulomb	23
Ohm's law	23
Units of heat and work, and their relation to the electrical units	24
Calculations	25
The Commercial Electrical Unit	31
The Watt	32
Questions	32

CHAPTER VI.

SOURCES OF ELECTRICITY—BATTERIES.

Theory of the voltaic cell	34
Amalgamated zinc	35
Smee's cell	36
Grove's cell	37
Daniell's cell	38
Latimer Clark's standard cell	39
Batteries of several cells	41
Questions	42

CHAPTER VII.

SOURCES OF ELECTRICITY—ELECTRO-MAGNETIC INDUCTION.

Electro-magnetic induction	43
Theory of electric generators	45
Lenz's law	47
Induction by variation of current	47
Dynamo machines	48
Questions	48

CHAPTER VIII.

MEASUREMENT OF ELECTRIC CURRENTS—GALVANOMETERS.

	PAGE
Galvanometers	50
The tangent galvanometer	50
Astatic needles	53
Reflecting galvanometers	55
Siemens' electro-dynamometer	58
Questions	61

CHAPTER IX.

MEASUREMENT OF ELECTRIC PRESSURE OR ELECTRO-MOTIVE FORCE.

Water analogies	62
Condenser method	64
Direct method	67
Metchim's electrometer	68
Small current method	69
Cardew's voltmeter	69
Questions	70

CHAPTER X.

EXPERIMENTAL MEASUREMENT OF RESISTANCES.

Wheatstone's bridge	71
———————— lecture model	71
———————— theory	72
Sliding bridge	75
Resistance coils	77
Bridge resistance box	77
Specific resistances	80
Galvanometer shunts	82
Questions	82

CHAPTER XI.

TELEGRAPHY.

Land telegraphy	86
Overhead wires	66
Underground wires	87
Single-needle instrument	87

viii   *Contents.*

	PAGE
Earth plates	88
Morse printing instrument	88
Morse code	89
Submarine telegraphy	90
Questions	94

CHAPTER XII.

THE TELEPHONE.

Graham Bell's first telephone	96
The Hughes microphone	97
The Gower-Bell telephone	98
Telephone exchanges	100
Questions	101

CHAPTER XIII.

ELECTRIC BELLS.

Electric bells	102
Electric bell indicators	103
Questions	104

CHAPTER XIV.

ELECTRIC LIGHTING.

Introductory—Theory of artificial lighting	105
Questions	107

CHAPTER XV.

ON THE CONVERSION OF ELECTRIC CURRENTS INTO HEAT.

Heat produced by friction of water	108
Electric current	109
Heat produced by electric current	110
Analogy between electric current and water current	111
Ampère's theory of magnetism	111
Theory of electric lamps	111
Use of carbon in electric lighting	112
Questions	114

Contents.

CHAPTER XVI.
INCANDESCENT LAMPS.

	PAGE
Swan incandescent lamps	115
Manufacture of Swan lamps	116
Efficiency of incandescent lamps	119
Questions	120

CHAPTER XVII.
ARC LAMPS.

Arc lamps	121
The Serrin lamp	122
The Crompton lamp	125
Carbons for arc lamps	128
Questions	132

CHAPTER XVIII.
DYNAMO MACHINES.

Direct and alternating currents	133
Alternating current machines	133
The Gordon machine	134
Direct **current** machines	137
The Gramme ring	137
The Burgin machine	137
The Siemens sub-type	140
Magnetization of the field magnets	140
Questions	142

CHAPTER XIX.
ELECTROLYSIS.

Description of the phenomenon	143
Laws of electrolysis	144
The voltameter	147
Electro-plating	148
Questions	149

CHAPTER XX.
ELECTRO-MAGNETS—DIAMAGNETISM AND MAGNE-CRYSTALLIC ACTION.

Electro-magnets	150
Torsion balance	152

	PAGE
Diamagnetism	153
Diamagnetic polarity	155
Magne-crystallic action	157
Effects of the surrounding medium	159
Questions	160

CHAPTER XXI.

THE INDUCTION COIL.

The induction coil	162
The condenser	165
Contact breakers	165
Mr. Spottiswoode's coil	166
Questions	167

CHAPTER XXII.

ON THE DISCHARGE OF THE INDUCTION COIL AND DISCHARGE GENERALLY.

Discharge	168
Secondary condenser	169
Induction coil and magneto-electric machine	171
Discharge in rarefied air	171
Striæ	176
Questions	178

CHAPTER XXIII.

ON THE NATURE OF MATTER.

The three states of matter	179
The size of molecules	180
Radiant matter	181
Crookes' experiments	182
Radiant matter exerts powerful phosphorogenic action where it strikes	182
Radiant matter proceeds in straight lines	186
Radiant matter, when intercepted by solid matter, casts a shadow	190
Radiant matter exerts strong mechanical action where it strikes	192
Radiant matter is deflected by a magnet	194
Radiant matter produces heat when its motion is arrested	199
Questions	200

CHAPTER XXIV.

ELECTRO-STATIC INDUCTION.

	PAGE
Charged **bodies**	201
Law of force	202
Induced charges	203
The Leyden jar	205
Residual charge	206
How strain is propagated	210
Induction precedes discharge	211
Lines of force are real	212
Questions	214

CHAPTER XXV.

SPECIFIC INDUCTIVE CAPACITY.

Definition	215
Faraday's experiments	216
Gordon's experiments	219
Specific inductive capacity **of gases**	225
Questions	225

CHAPTER XXVI.

ANALOGIES BETWEEN THE MUTUAL ACTIONS OF CURRENTS, OF MAGNETS, AND OF ELECTRIFIED BODIES, AND THOSE OF BODIES PULSATING AND VIBRATING IN FLUIDS.

Bjerknes' researches	227
Phases	229
Attraction of light bodies and soft iron	229
Attraction or repulsion of a compass needle	229
Two magnets of unequal strength	229
Diamagnetism	229
Illustrations of phases	230, 231
Diamagnetism (*continued*)	232
Lines of force	233
Stroh's experiments	234
Questions	235

CHAPTER XXVII.

ACTION OF MAGNETISM AND ELECTRICITY ON POLARIZED LIGHT.

Polarized light	236
Natural rotation	237

	PAGE
Faraday's discovery of magnetic rotation	237
Difference between magnetic and natural rotation	238
Faraday's paper	239
Direction of the rotation	240
Action of static electricity on light—Kerr	243
Questions	243

CHAPTER XXVIII.

CLERK MAXWELL'S ELECTRO-MAGNETIC THEORY OF LIGHT.

Maxwell's theory	245
Comparison of velocities	247
Gordon's experiments	248
Gibson and Barclay's experiments	248
Boltzmann's experiments	249
Crystalline sulphur	249
Schiller's experiments	249
Silow's experiments	250
Boltzmann's comparison for gases	250
General conclusion	251
Questions	251

EXAMINATION PAPERS.

NO.	
I.	252
II.	252
III.	253
IV.	254
V.	254
VI.	255
VII.	256
VIII.	256
IX.	257
X.	258

INDEX	259

SCHOOL ELECTRICITY.

CHAPTER I.

ELECTRIC CURRENTS.

For all practical purposes we may treat electricity as if it were a material incompressible fluid, which, when in motion, produces certain effects which we can observe.

1. Conductors and insulators. Certain substances through which electricity can flow freely are called *conductors*, and certain others through which it cannot flow freely are called *insulators*. No substances are quite perfect either way, i.e. the best conductors offer a certain *resistance* to the flow of electricity, and the best insulators allow a certain minute flow through them.

For all practical purposes copper and other metals and carbon may be considered as conductors, and gutta-percha, glass, and dry air may be considered as insulators. The best insulators offer many million times the resistance to the flow of electricity that is offered by conductors.

The fact that the best conductors of electricity, such as copper and iron, are metals, and can be formed into wires, enables electric currents to be conveyed to great distances.

Upon this fact depend all the practical applications of electricity, such as telegraphy and electric lighting.

We cannot tell by looking at a wire whether a current is flowing in it or not, but there are numerous and certain methods of detecting not only the fact, but *in which direction* the current is flowing, and *how much* current is flowing at any moment.

2. Effects of the electric current.

The principal effects are,—

(*a*) **Magnetic effects.** Electricity and magnetism are very closely allied, but we must postpone the consideration of their relations till we have said something about magnets.

Telegraphy depends on the fact that an electric current will cause a pivotted magnet, such as a compass needle, to move to the right or left according to the direction of the current. An operator at one end of the wire sends currents sometimes in one direction, sometimes in the other, causing corresponding motions of a needle at the other end. It is agreed that certain motions shall represent certain letters of the alphabet. For instance, two motions to the right might represent a, one to the right and one to the left b, and so on.

(*b*) **Heating effects.** Electric currents heat the substances through which they pass, the amount of heat depending on certain definite laws, which we shall presently discuss.

Electric lighting depends on the concentration of the heat of an electric current in one or more spots, where it is used to heat a substance such as carbon to whiteness and to give light.

(*c*) **Chemical effects.** When an electric current is passed through certain solutions containing metals in chemical combination it decomposes them, liberating the metal at one place and the other constituent of the compound at another.

On this fact depends the industry of **electroplating**. For instance, if a copper fork or spoon is placed in a solution of potassio-cyanide of silver, and an electric current is sent through the fork, and from it into the solution, silver is deposited all over the fork, and gives it the appearance well known to us as "silver plate."

(*d*) **Effects on the human body.** A certain class of electric currents cause muscular twitching when sent through the body, and produce the well-known sensation of the *electric shock*.

(*e*) **Sparks.** When the conductor carrying a current is interrupted, and sufficient force is applied to the current to drive it across the intervening air or other insulator, it bursts through in the form of a spark. The largest electric sparks known are produced in Nature as "flashes of lightning."

(*f*) **Effects on polarized light.** When rays of light are placed under certain conditions they can be acted on by electricity. This action will be discussed towards the end of this book.

Questions on Chapter I.

1. What is a conductor?
2. What is an insulator?
3. Give instances of each.
4. What are the principal effects of the electric current?

CHAPTER II.

LAWS GOVERNING THE FLOW OF ELECTRIC CURRENTS—OHM'S LAW.

3. Ohm's law. The strength of an electric current depends on the force driving or moving it, called the *electro-motive force* or *electric pressure*, and on the *resistance* through which it has to be driven.

If the electro-motive force is increased, the current is proportionately increased; and if the resistance is increased, the current is proportionately *diminished*. This is expressed in somewhat more technical language by saying that the current is directly proportional to the electro-motive force, and inversely proportional to the resistance.

The above law was discovered by a German professor named Ohm, and is consequently known as **Ohm's law**. It is the foundation of all electrical measurement, and must therefore be carefully remembered and studied.

If we double the electro-motive force (keeping the resistance constant), we double the current.

If we double the resistance (keeping the electro-motive force constant), we halve the current.

If we double both the electro-motive force and the resistance, we keep the current unaltered.

Thus we may express Ohm's law by saying that the current equals the electro-motive force divided by the resistance; which may be written,—

$$\text{Current} = \frac{\text{Electro-motive force.}}{\text{Resistance.}}$$

For convenience in writing we agree to use the following abbreviations,—

C stands for Current.
E „ „ Electro-motive force or electric pressure.
R „ „ Resistance.

Ohm's law may then be written,—

$$C = \frac{E}{R}$$

We shall return to the study of Ohm's law in a later portion of this book.*

4. Speed of electric currents. When an electric current is admitted into a wire its effects are felt at the other end after an unappreciably short fraction of time.

It has therefore been assumed that electricity must travel with great velocity in a wire.

This is not necessarily the case, as the following illustration will show.

Suppose a metal tube, a mile long, to be quite full of water, and to be connected at one end to a force-pump, which would force water very slowly along it, say at the rate of a foot per minute.

The instant that the force-pump commenced working a current would start, and water would begin to flow out of the far end of the tube, and could there turn a small water-wheel, or do other work.

Here, although the velocity of the flow is very small, the commencement of the current at the far end has been practically simultaneous with that at the near end.

We see then that the speed with which electric signals (which are given by the starting and stopping of currents) are transmitted gives no information as to the speed of the current itself, which is a matter of which we are profoundly ignorant.

If the water-tube, instead of being metal, were made of india-rubber, or some such elastic material, the starting of the force-pump at one end would not be felt instantaneously at the other, as the first effect would be to stretch the parts of the tube near the force-pump. The flow would not commence at the far end until these had had time to contract again.

* See page 23.

An analogous phenomenon is observed in electrical conductors which, as in submarine cables, are covered with an electrically elastic material, such as gutta-percha. The first effect of starting the current is to electrically stretch or "charge" the gutta-percha, and not till it has recovered from its stretching is the current perceived at the far end.

The practical result is that with submarine cables a number of special methods have to be used to telegraph quickly, as signals dependent on the ordinary starting and stopping of currents would be too slow to enable the cable to earn a reasonable revenue.

QUESTIONS ON CHAPTER II.

1. State Ohm's law.
2. If a given electric pressure produces a given current through a given resistance, what will be the effect on the current of

(a) Doubling the electric pressure without altering the resistance?

(b) Doubling the resistance without altering the pressure?

(c) Doubling both the resistance and the pressure?

3. Does the fact that electric signals are transmitted with great rapidity through great lengths of wire give us any information as to the speed of electric currents?

4. Would it be possible to transmit signals quickly along a pipe by means of a slow moving current of water; if so, how?

CHAPTER III.

MAGNETS AND ELECTRO-MAGNETS. LINES OF FORCE.

5. Magnets. We have said that an electric current has certain effects on magnets. We have now to consider what a magnet is.

A bar of **steel** when *magnetized*, possesses certain **properties**.

(1) It attracts **soft** iron.
(2) It attracts or repels other magnets.
(3) When pivotted **or** suspended so that it can turn freely, one end points towards the north, and the other end towards the south.

Thus **the two ends are** *different.*

The **ends of a** magnet **are called its " poles."** That end which points northward **is** called **the north pole.**

Similar **poles of magnets** *repel* **each other.** Like poles attract. Thus, if we suspend or pivot one magnet (as, for instance, a compass needle), **and bring the** north pole **of** another magnet **near** its north **pole, the latter** will be repelled; but if the south pole **of the magnet be** used, the north pole of the compass will be **attracted.**

Magnetic induction. **If a piece of** soft **iron be** brought

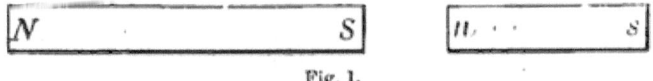

Fig. 1.

near **a magnet it will become a magnet temporarily,** i.e. as long as it remains near the steel magnet. The magnet pole will induce a pole **of opposite** name to itself on the end of the soft iron **nearest to it. It** will then attract the **pole** which it has induced, or, in other words, it will attract the soft iron.

6. Action between magnets and currents. Magnets and electric currents are very closely related to each other.

Action of currents on magnets. Let a compass needle be suspended either on a point, or by a thread, and let a straight wire be brought near it, parallel to it, and above it (fig. 2). Let now a current be sent through this

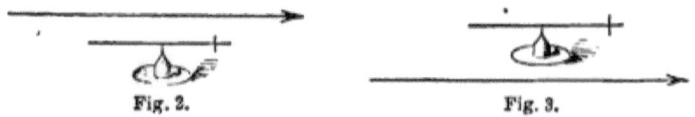

Fig. 2.   Fig. 3.

wire. *The needle will be deflected, i.e. it will take up a position making a certain angle with the wire, which angle will increase as the strength of the current increases.* If the direction of the current be reversed the direction of the deflection will be reversed. If the direction of the current be kept the same, and the wire placed below (fig. 3) instead of above the needle, the deflection will be reversed. If the direction of the current be reversed, *and* the wire placed below the needle, the deflection will be twice reversed, i.e. it will remain the same.

If now the wire be bent round the needle (fig. 4), the action of the bottom piece will cause the same deflection as that of the top one, for it is below the needle, and the current

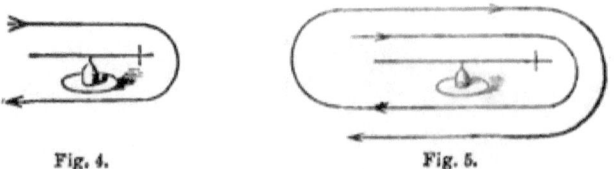

Fig. 4.   Fig. 5.

is in the opposite direction, and so the action on the needle is greater than that of either portion singly. It may be bent again and again (fig. 5), and the effect is always to increase the deflection.

A current in a certain direction always deflects the marked

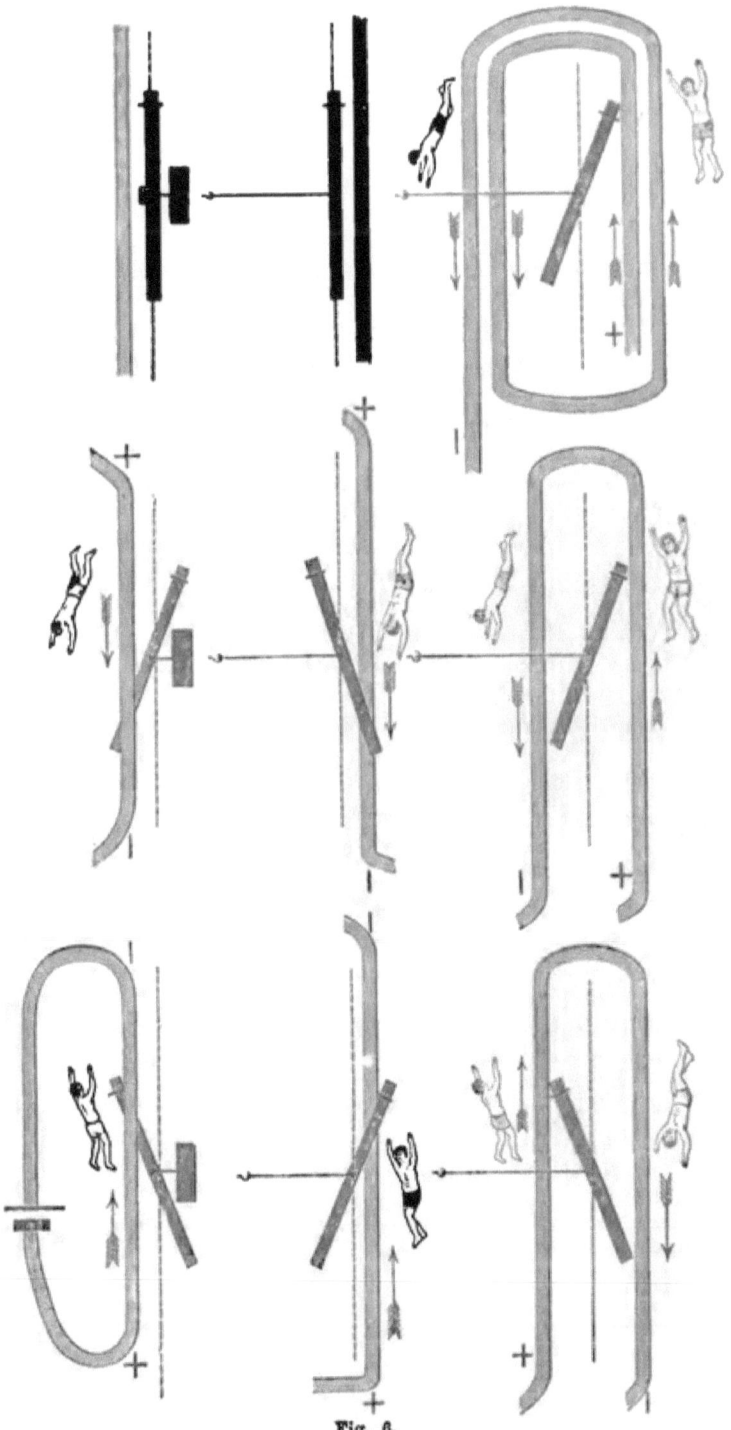

Fig. 6.

end of the needle in the same direction. We will see which direction this is.

Experiment will show us that if we imagine a little man swimming in the current with his face always turned to the needle, the north-pointing end of the needle will always be deflected to his left hand.

An examination of the diagrams (fig. 6) will make this clearer than any explanation.

Action of magnets on currents. If the magnet be fixed and the wire carrying the current be so suspended that it can move freely, it will be deflected by the magnet. We can find the directions of motion by remembering that the attractive and repulsive forces being the same as before the motion of the wire, will therefore be always in the opposite direction to the former motion of the magnet.

This can be made clear by taking two chairs and letting a boy kneel on one and placing the other near him. Let the one he is kneeling in represent the wire carrying the current, and the other the magnet. Now if the other boy hold the "current-chair" still, and the boy in it pushes the "magnet-chair," the latter will move forward, while if the "magnet-chair" is held still, the effect of the boy in the "current-chair" pushing will be to move the "current-chair" backwards.

7. Electro-magnets. If a wire be wound spirally round an iron bar, and a current be sent through the wire, the bar will become a magnet, and remain magnetized as long as the current continues to flow.

The direction of the poles of the magnet depends on the direction of the currents, and will be reversed when the current is reversed.

If we imagine a little man to be swimming in the current down stream, with his face towards the magnet, the north pole will be on his left hand.*

---

* Boys who have access to a swimming bath, or other bathing-place, may realize this method of remembering polarity by the following experiment: Two boys may hold a broom-stick horizontally about two feet below the surface of the water; a third may then swim over it, and as soon as he has passed it, dive under it, and passing under it

Electro-magnets—Telegraphy.

A magnet such as we have described, which only exists while an electric current is flowing, is called an *electro-magnet*. Electro-magnets can be made much more powerful than permanent magnets of the same size. Fig. 7 represents an electro-magnet.

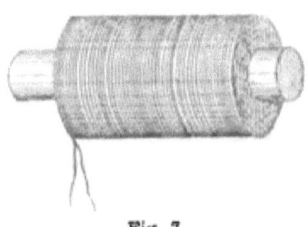

Fig. 7.

8. Telegraphy. All telegra-

come to the surface, face uppermost, on the side he started from. He will thus have been looking towards the broom-stick during the whole, both of the swim and the dive, and the *same end of it will always have been on his left hand*. This is the end which, if the stick had been an iron bar and the swimmer had been swimming down an electric current, would have become a north pole.

Similar experiments may be used for illustrating the actions of currents on magnets, shown on p. 9. A broom-stick, one end marked "North," may be held horizontally under water by two cords attached near its centre to a weight at the bottom of the bath (fig. 8). Corks attached to it will keep the cords tight. If the experimenter then—

Fig. 8.

phic instruments depend on one or other of the actions of currents on magnets which we have just described.

We see that if a coil of wire is placed round a compass needle, as in fig. 5, then, by sending a current sometimes in one direction and sometimes in the other through the wire, we can deflect the needle to right and left at will.

Also, if an iron lever be hinged near an electro-magnet and pulled away from it by a spring, we can, by starting and stopping a current in the coils of the magnet, cause the lever to be moved backwards and forwards at will, and so actuate various apparatus.

Now the place where the current is started, stopped, and reversed may be many hundreds of miles from the place where the needle or electro-magnet is, provided only that the two places are connected by a wire.

Thus the person at the "sending station" can, by sending, stopping, and reversing currents, cause any succession of movements that he pleases of the pivotted needle, or magnet lever, at the receiving station, and it is previously agreed that certain movements should represent certain letters of the alphabet.

Some few of the elaborate and beautiful instruments which have been devised, for practically carrying out the principles which we have here indicated, will be described in a later part of this book.

9. Lines of force. The power of a magnet (whether a permanent magnet or an electro-magnet) extends to a considerable distance from its poles. For instance, if an iron bar is held some two feet from the pole of a large magnet, such as is used in dynamo-machines, and moved about, it feels as if it was being moved through thick treacle. The direction of this power or force follows certain lines, called *lines of force*.

The following is a method of tracing out the directions of these lines of force. A straight steel magnet may be used.

always looking at the stick—swims over it and dives under it, moving *parallel* to the stick, and whenever he passes either over or under the "North" end, pushes it to his left, he will find that he always pushes it in the same direction for the same direction of swimming.

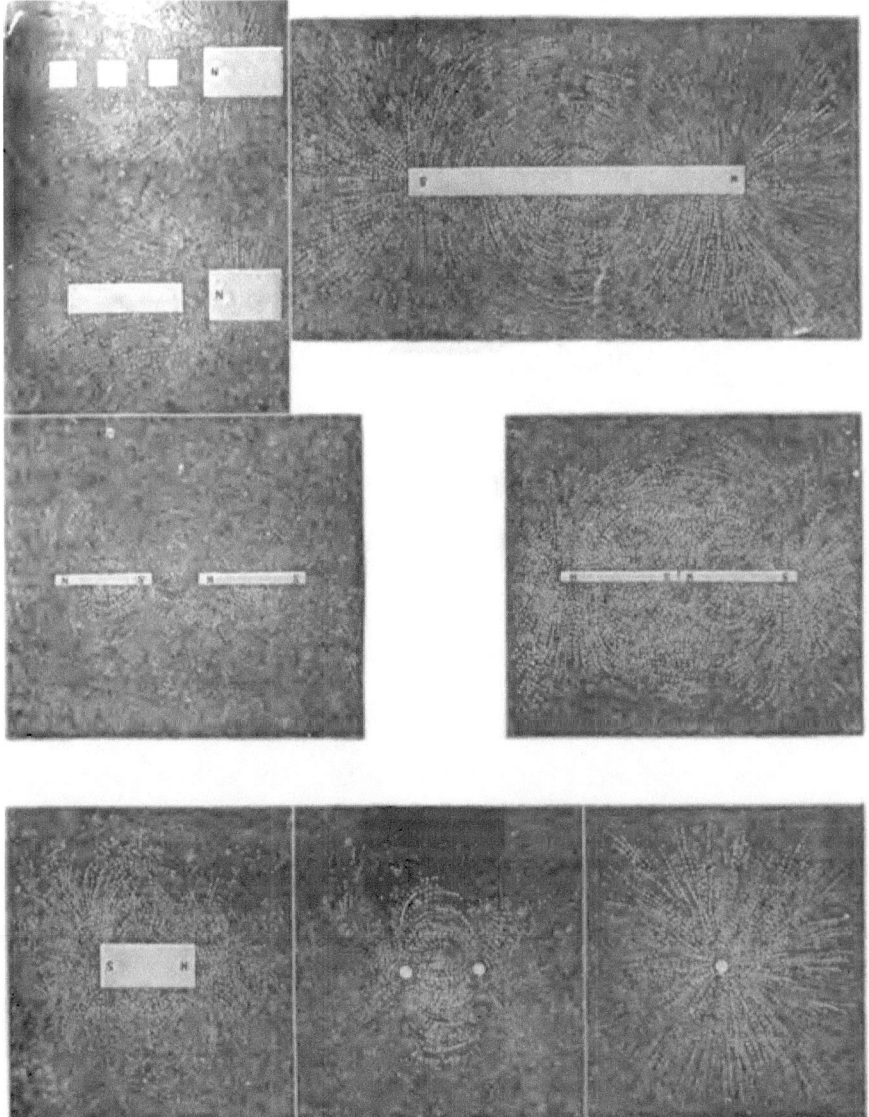

Fig. 9.

Take a board of the same or greater thickness than the magnet, and let the magnet into it so that its upper side is flush with the upper surface of the board.

Lay a piece of smooth paper over all and fix it.

Now with a sieve or "colander" dust fine iron filings all over the paper.

As each falls near the magnet, it is magnetized by induction and turns round so as to lay its longest diameter in the line of force at that point.

Each of these minute magnets attracts the next till a continuous chain is formed all along each of the lines of force. The board must be tapped from time to time to overcome the friction of the filings on the paper.

If it be desired to preserve the curves, a piece of card with its under side gummed may be laid upon them. When the gum is dry, the filings stick to the card. The magnet may be placed under a piece of glass instead of paper if preferred, in which case it will not be necessary to let it into a board.

A good way to prepare these curves for exhibition in a magic lantern is to coat a piece of glass with some transparent cement,* which melts on being heated. When it is quite hard, a magnet must be placed under the glass, and the filings dusted on. The glass must then be carefully carried to an oven, warmed till the cement is soft, when the filings will sink in. On removing the glass and allowing it to get cold, the filings will be all fixed in position.

Fig. 9 is an engraving of the actual lines of force determined by Faraday for different magnets.

10. Yokes and magnetic circuits. If two bar magnets are laid near each other, as in fig. 10, a piece of iron I will be attracted with a certain force, and it would be attracted with equal force at the other end of the magnets. If now a piece of iron Y be placed across the other ends of the magnets, as in fig. 11, the iron I will be attracted with rather more than double the force it was formerly attracted with, but

---

* The cement sold under the name of "Mend all" answers very well. It may be diluted with water for the purpose.

no attraction at all would be exercised on it if it were placed near the other end beyond the cross-piece Y.

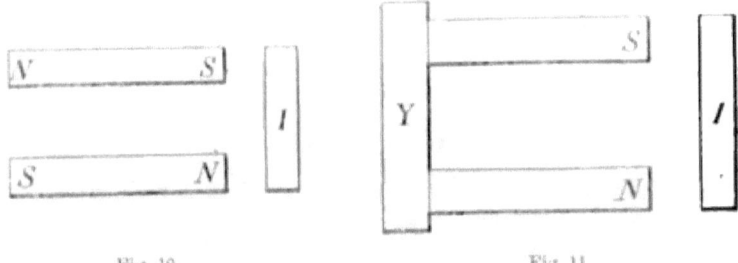

Fig. 10.   Fig. 11

The effect of the cross-piece appears to be to make the lower poles disappear, but to add their strength to the upper poles which remain.

The cross-piece Y is commonly called the *yoke* of the magnet.

If in the experiment of fig. 11 the iron I be allowed to come upon the poles N, S, so as *to complete the magnetic circuit,* all external magnetic effects disappear, i.e. the magnets will no longer attract iron filings, but on the other hand, the force required to pull off either the bar or the yoke is very much greater than it would be with any arrangement of the magnets where the magnetic circuit is open, as for instance in fig. 10.

We see that by *closing the magnetic circuit* we cause all the *lines of force* that were previously wandering about in the neighbourhood of the magnets to leave the external space, and concentrate themselves within the magnets. A place through which lines of force pass is called a *field of force.* The more concentrated the lines the stronger the field.

The study of the disposition of the lines of force, and the best way to concentrate them in any desired place is of great practical importance, as *dynamo-machines,* which are the machines used for generating the powerful currents used for electric lighting, depend for their action on the motion of wires through *fields of force.* The better these *fields of force* are arranged for the concentration of the lines of force just where they are wanted, the more lamps will a machine of given size and cost be able to maintain.

Questions on Chapter III.

1. A compass needle is suspended so that it points to the north. A wire is placed over it and parallel to it, and a current sent through the wire from S. to N. Will the N. pointing end of the needle move to the E. or the W.?

2. The same wire with the same current is placed under the needle; in which direction will the N. pointing end move?

3. The wire is bent round so that the current flows from S. to N. over the needle, and from N. to S. under it.
 Will the N. pointing end of the needle move E. or W.?
 Will it move more or less than before?

4. What is an electro-magnet? Give a sketch, and describe?

5. How can electro-magnets be used in telegraphy?

6. What is meant by the lines of force of a magnet? How may they be observed?

7. What is the effect of connecting one pair of poles of two bar magnets by a "yoke," the magnets being laid so that the N. pole of one points in the same direction as the S. pole of the other?

CHAPTER IV.

RELATION OF ELECTRIC CURRENTS AND ELECTRIC PRESSURE TO MECHANICAL FORCES.

11. THE work which an ordinary water-wheel can do depends upon two things—upon the quantity of water passing per minute, and upon the height through which it falls.

We cannot say what horse-power a water-wheel worked by water, falling through so many feet, can develop, unless we know how much water passes in each minute; neither can we say what horse-power the wheel can develop, with a stream of so many pounds' weight of water per minute, unless we know through what height the water is falling.

As soon, however, as we know both the height and the quantity we can (after making a proper allowance for the waste caused by imperfect construction of the wheel) calculate exactly the horse-power which will be developed.

If we have a greater fall we can get our required H.P.* with proportionately less water. If we have more water we can get the H.P. with proportionately less fall.

If we double the fall we halve the quantity of water required; if we double the quantity of water we halve the fall.

Thus in low-lying districts, such as Surrey, Kent, or Cambridgeshire, where the rivers are wide and sluggish, but the falls small, wide wheels of small diameter are used; but in mountainous districts, such as Cumberland or Lancashire, where there is not much water, but where it runs down high hills, narrow wheels of large diameter are used; these use but little water, but by their large diameter allow it to exercise great " leverage."

* H.P. stands for horse-power.

C

12. Foot-pounds. Thus the work which any quantity of water or other heavy substance can do on falling through a certain distance *is measured by the weight in pounds multiplied by the height in feet*, and is called the work in *foot-pounds*. To raise the same quantity of water to the same height requires the same expenditure of work as it gave out in falling.

Thus to carry a bucket of water, which when full weighs 60 lbs., to the top of a 10-foot ladder requires an expenditure of (60 × 10 =) 600 foot-pounds of work.

The cistern of an ordinary shower-bath is about 8 feet above the bath, and holds about 7 gallons or 70 lbs.* of water. The servant who pumps up the water must therefore (in addition to the work due to the stiffness of the pump) expend 8 × 70 = 560 foot-pounds of work in filling the cistern.

13. Rate of doing work.—Horse-power. Now it is obvious that it will require a stronger man to fill the shower-bath cistern once every five minutes than to fill it once every ten minutes, and that if we wanted it filled every ten seconds the work would be more than any man could do, but that it could be done if we could arrange to have a horse to work the pump.

We see that *the faster we do our work, or the more foot-pounds of work that we do in each minute*, the more *power* we must exert.

Power, or the *rate of doing work*, is commonly measured in "horse-power." That is, it is assumed that a horse can exert a certain number of foot-pounds in each minute. A steam-engine that can do just that number of foot-pounds in a minute is said to be of one horse-power; one that can do twice that number, of two horse-power, and so on.

A horse-power is assumed to be equal to 33,000 foot-pounds per minute. That is, an engine of one horse-power would take one minute to raise 1000 lbs. of water 33 feet.

This number is fixed, and adopted by all engineers. It represents, however, more work than any one horse could do.

The way it was arrived at was the following. When James

* A gallon of water weighs 10 lbs.

Watt first **began** to sell steam-engines he found that his customers, **who** had been accustomed to use horses, used to order engines to do the work of five or six or seven horses, as the case might be. In order **to** determine how much **a** horse could do, he arranged **a** pulley at the edge of a deep well, over which he hung a rope. He hung a heavy weight in the well, and harnessed a strong horse to the other end of the rope.

He then noted the weight in pounds and the distance **in** feet which the horse raising it could **walk** in a minute. From this experiment he found that **a** strong horse could do just **22,000** foot-pounds of work **in** a minute.

Then being an honest man, and anxious that **his customers** should **not have** any cause **of** complaint against him, he added **50** per **cent.**, and fixed the "horse-power" at 33,000 foot-pounds per minute, **so that** when he sold an engine as of **one** horse-power it could really do as much work as **a** horse and a half, and **so on**.

14. Electrical horse-power. To produce **a** certain electrical effect a certain number of foot-pounds of mechanical energy **must** be expended.

To **continuously** produce a certain electrical effect, minute by minute, **a certain number** of foot-pounds per minute must be expended, **that is, a certain horse-power** must be expended.

We **cannot say that a certain current is** equal to a certain horse-power, any **more than we can** say that a certain current of water is equal **to** a certain horse-power when **we** do not know the fall.

Neither can we say that a certain electric pressure is equal **to** a certain horse-power when we do not know the current, any more **than we can** say that **a certain head** of water **is** equal **to a certain horse-power, when** we **do** not know **the** number of pounds **of water per** minute which are flowing.

To know **the horse-power of a** water-wheel **we must know** *both* the pounds of **water** which flow per minute and the height of fall, and **we** know that the horse-power is proportional to the product **of** these two **quantities**.

To know the horse-power of an electrical effect, for instance the horse-power which **is** being expended in an electric lamp, we must know *both* the current and the electric pressure, and, as in the water analogy, the horse-power is proportional to the product of these two quantities.

A certain definite quantity of electrical energy is equal to a certain definite quantity of mechanical energy.

In order to know how much electrical energy is equal to a certain mechanical energy, we must know in what way both energies are measured.*

We already know that mechanical energy is measured in foot-pounds per minute, and in the next chapter we shall study "electrical units."

Questions on Chapter IV.

1. If we have two water-wheels, one with a fall of 12 feet and one with a fall of 4 feet, what are the relative quantities of water we must use so that each wheel may develop the same horse-power?

2. What is meant by "foot-pounds"?

3. How many foot-pounds of work are required to raise 30 lbs. through 12 feet?

4. How many to raise one ton through 6 inches?

5. How many to raise one cwt. 20 yards?

6. What is a "horse-power"?

7. How many H.P. are required to raise 1000 lbs. 66 feet in (a) one minute, (b) half a minute, (c) two minutes?

8. How many to raise 2000 lbs. 33 feet in the same times?

9. How many foot-pounds per second will be developed by one H.P.

---

* Up to this point in our studies we have learnt that a certain quantity of electrical energy is equal to a certain quantity of mechanical energy, but we do not know the units in which the former is measured.

Our position is that of an Englishman ignorant of French measures, who, if he was offered six kilogrammes of potatoes for a shilling, would know that potatoes were being sold at a definite price, but would have no practical knowledge of what that price might be until he had studied the French system of measures, and found how large a basket of potatoes weighed a kilogramme.

CHAPTER V.

ELECTRICAL UNITS AND THEIR RELATION TO EACH OTHER, AND TO THE HEAT AND WORK UNITS.

This chapter looks difficult, but there is nothing in it beyond the capacity of a boy with a knowledge of arithmetic. The only real difficulty lies in the terms used not being familiar ones. This can only be overcome by constant practice. One or two problems similar to those in this chapter should be set daily, and the class exercised in them, till Ohm's law in every possible form is as familiar as the multiplication table. It is not necessary for a class to master this chapter altogether before going on with the book. After reading it once they may go on with the rest of the book, but should return daily to some exercises on this chapter.

In order to force water through a pipe offering resistance to the flow, a certain **pressure** or "aqua-motive force" must be supplied by a force-pump or otherwise. Similarly, as we have already stated, in order to force a current of electricity through a wire, a certain *electric pressure* or *electro-motive force* must be supplied by the generator. A small electro-motive force will force a small *current* through a given *resistance*; a larger, a proportionately larger one.

15. Current.—Ampère. The unit of electric current is called the *Ampère*. It is a unit of flow or of stream. We speak of an electric current of so many ampères in the same way as we might speak of a water current of so many gallons per minute.* A 20-candle Swan † lamp (new pattern) may take a current of about ·7 ampère.

16. Pressure.—Volt. The unit of electric pressure is called the *Volt*. It is analogous to steam-pressure or to head of water. We speak of an electric pressure of so many volts as we might speak of a steam-pressure of so many pounds to the square inch, or a head of water of so many feet. One Daniell's † cell gives a pressure of nearly one volt. One 20-candle Swan lamp (new pattern) may require a pressure of about 100 volts.

17. Resistance.—Ohm. We know that generally a pipe of small bore offers a greater resistance to a flow of water than one of large bore; but the relation between the resistances of different pipes follows extremely complicated laws. We know, for instance, that a pipe of 1 square inch bore offers more resistance than one of 2 square inches, but we cannot say that it offers exactly, or even approximately, twice the resistance.

Electric resistance, however, as we have already stated,‡ follows a very simple law. For a wire of a given substance it is directly proportional to the **length**, and inversely proportional to the **cross section**. Thus, if we double the length of a wire we double its resistance. If we double the section we halve its resistance. If we double the diameter we quadruple the section and reduce the resistance to one quarter. If we double the length and double the section the resistance remains unaltered.

The unit of electrical resistance is called the *Ohm*. It is defined as the resistance at a temperature of 0 C., of a column of mercury **of one** square millimetre section and **of a certain length**. This length **was chosen** so that **the** value of the practical ohm might as nearly as possible equal that deduced from measurements of the absolute electro-magnetic unit §.

A set of such measurements were made in 1862, by a committee of the British Association; and the **result** of their

---

* In the case of the water-flow we have no one word to express the **strength** of the stream, but have to speak of quantity per time.
† These will be described later on.   ‡ Page 4.
§ For the theory of the determination of the ohm, see my "Electricity," 2nd edition, vol. i. p. 303.

determination is that the standard mercury column has a length of, as nearly as possible, 104 centimetres. This is the value which is at present in practical use. It may be called the B. A. Ohm.

Recent investigations have, however, shown that it is about 1 per cent. too low; and at the Congress of Electricians, which met in Paris on Sept. 15, 1881, an International Commission was appointed to re-determine it with all possible accuracy. Whatever value the Commission arrives at is to be called the " Paris Congress Ohm," and is to be adopted permanently, and is not to be again changed, even if a further re-determination should show that it is not perfectly accurate.

To get some idea of the magnitude of the ohm, we may note that a mile of No. 16 copper bell-wire has a resistance of about 14 ohms, while the Atlantic cable has a resistance of about 36,000 ohms. The carbon thread of a 20-candle Swan incandescent lamp (new pattern) has, when hot, about 150 ohms resistance, while the resistance of the heated air in an electric arc varies from 6 ohms to 1 ohm.

18. Unit of quantity.—Coulomb. The unit of electrical quantity is called the *Coulomb*, and we can speak of a current as one that conveys so many coulombs per second through the wire. A current of a strength of one ampère conveys one coulomb per second. This quantity is not much used in practice.

19. Ohm's law. The three units, volt, ohm, ampère, are connected by what is known as *Ohm's law*.

As we have already said (page 4), Ohm's law states that the *current in any circuit is directly proportional to the electromotive force, and inversely proportional to the resistance,* and the units are so chosen that when there is one ohm resistance in circuit an electro-motive force of one volt produces a current of one ampère.

We see then that two volts acting through one ohm would give two ampères, or one volt acting through two ohms would give ½ an ampère.

Ohm's law may thus be written,—

$$\text{Current in ampères} = \frac{\text{Electro-motive force in volts.}}{\text{Resistance in ohms.}}$$

This is commonly abbreviated into the form,—

$$C = \frac{E}{R} \quad . \quad . \quad . \quad . \quad (1)$$

This may also be written,—

$$E = C R \quad . \quad . \quad . \quad (2)$$

or,—

$$R = \frac{E}{C} \quad . \quad . \quad . \quad . \quad (3)$$

By the use of these formulæ we can solve problems such as the following, which occur daily in electric engineering.

(1) A machine gives an electric pressure of 60 volts. What current will it send through a resistance of 5 ohms?

We have from (1),—

$$C = \frac{60}{5} = 12 \text{ ampères.}$$

(2) What electric pressure must a machine have to send a current of 2 ampères through a resistance of 25 ohms?

We have from (2),—

$$E = 2 \times 25 = 50 \text{ volts.}$$

(3) What is the resistance of a circuit when an electric pressure of 800 volts sends a current of 10 ampères through it?

We have from (3),—

$$R = \frac{800}{10} = 80 \text{ ohms.}$$

20. Units of heat and work, and their relation to the electrical units.—Energy and horse-power. *The rate at which energy is being expended*, as for instance in maintaining a current, is, as we have already stated, commonly measured in "horse-power."

One horse-power is equal to 33,000 foot-pounds per minute, i.e. can raise 33,000 lbs. 1 ft. per minute, or 1 lb. 33,000 ft. or 100 lbs. 330 ft. in the same time.

When horse-power is being expended in sending a current through a resistance, the conductor offering the resistance is heated. The quantity of heat produced per minute is equal to the heat which must be expended per minute in maintaining the current.

The horse-power required to maintain a current is, other things being equal, proportional to the square of the current. Thus, if one H.P. could maintain one ampère through a given resistance, 4 H.P. would be required to maintain 2 ampères through the same resistance.

The heat produced in the conductor is proportional to the square of the current. Thus, 2 ampères will produce four times as much heat in a certain wire as one will.

The horse-power required to maintain a certain current through a resistance is proportional to the resistance, and the heat produced by the current is proportional to the resistance.

(*Corollary*. The heat produced per unit of length in a wire, on which depends the temperature to which the wire will be raised, is proportional to the resistance per unit of length.)

The horse-power required to maintain a certain current under a certain pressure, is proportional to the current multiplied by the pressure.

CALCULATIONS.

21. Relation between horse-power, current, and resistance. One horse-power can maintain a current of one ampère through 746 ohms. Or one of two ampères through,—

$$\frac{1}{2^2} = \frac{1}{4}, \text{ of 746 ohms, &c.}$$

This is expressed generally by saying that the horse-power required to maintain a current is $\frac{1}{746}$ part of the square of the current in ampères multiplied by the resistance in ohms. This is abbreviated as follows,—

$$\text{H.P.} = \frac{C^2 R}{746} \qquad . \qquad . \qquad . \qquad (4)$$

This may also be written,—

$$C^2 = \frac{746 \text{ H.P.}}{R} \qquad . \qquad . \qquad . \qquad 5)$$

or,—
$$R = \frac{746 \text{ H.P.}}{C^2} \qquad . \qquad . \qquad . \qquad . \qquad (6)$$

Problem (1).

What horse-power is required to maintain a current of 10 ampères through a resistance of 6 ohms?

We have from (4),—
$$\text{H.P.} = \frac{10 \times 10 \times 6}{746} = \frac{600}{746}$$

or a little less than $\frac{6}{7}$ of a horse-power.

(2) What current can 16 H.P. maintain through a resistance of 64 ohms?

We have from (5),—
$$C^2 = \frac{746 \times 16}{64} = 186 \cdot 5.$$

whence $C = 13 \cdot 65$ ampères.

(3) Through what resistance can 10 H.P. maintain a current of 2 ampères?

We have from (6),—
$$R = \frac{746 \times 10}{2 \times 2} = 1865 \text{ ohms.}$$

22. Relation between horse-power, current, and pressure. Equation (4) shows us that,—
$$\text{H.P.} = \frac{C^2 R}{746} \qquad . \qquad . \qquad . \qquad . \qquad (4)$$

Equation (3) shows us that,—
$$R = \frac{E}{C} \qquad . \qquad . \qquad . \qquad . \qquad . \qquad (5)$$

Inserting in (4) the value of R given by (3) we have,—
$$\text{H.P.} = \frac{\frac{C^2 E}{C}}{746} = \frac{E C}{746} \qquad . \qquad . \qquad . \qquad (7)$$

or the horse-power expended in sending a current through any resistance, constant or variable, is $\frac{1}{746}$ part of the current in ampères multiplied by the pressure in volts which is driving it.

Equation (7) may also be written,—

$$E = \frac{746 \text{ H.P.}}{C} \quad . \quad . \quad . \quad . \quad (8)$$

or,—

$$C = \frac{746 \text{ H.P.}}{E} \quad . \quad . \quad . \quad . \quad (9)$$

Problem (1).

How much heat will be developed in a circuit by a current of 18 ampères driven by **an** E.M.F. of 200 volts?

We **have** from (7),—

$$\text{H.P.} = \frac{18 \times 200}{746} = 4\cdot 82 \text{ horse-power.}$$

(2) What pressure must a machine have in order that 5 H.P. may just maintain **a** current of 25 ampères in the circuit.

We have **from (8),**—

$$E = \frac{746 \times 5}{25} = 149\cdot 2 \text{ volts.}$$

(3) A machine has a pressure of 60 volts, what **current** will be developed by 80 H.P.?

We have **from (9),**—

$$C = \frac{746 \times 80}{60} = 994 \text{ ampères.}$$

23. Relation between horse-power resistance and pressure.

Equation (7) gives us,—

$$\text{H.P.} = \frac{E C}{746} \quad . \quad . \quad . \quad . \quad (7)$$

Equation (1) **gives us,**—

$$C = \frac{E}{R} \quad . \quad . \quad . \quad . \quad (1)$$

Substituting in (7) the value of C given by (1) we have,—

$$\text{H.P.} = \frac{E \frac{E}{R}}{746} = \frac{E^2}{746\, R} \quad . \quad . \quad . \quad (10)$$

This may also be written,—

$$E^2 = \text{H.P.} \cdot 746 \, R \quad \ldots \quad (11)$$

or,—

$$R = \frac{E^2}{746 \, \text{H.P.}} \quad \ldots \quad (12)$$

Problem (1).

What H.P. is expended by a pressure of 99 volts working through a resistance of 140·5 ohms?

We have from (10),—

$$\text{H.P.} = \frac{99 \times 99}{746 \times 140 \cdot 5} = \cdot 0936.$$

(2) What pressure will be developed if $\frac{1}{10}$ of a horse-power is employed in sending a current through 30 ohms?

We have from (11),—

$$E^2 = \frac{1}{10} 746 \times 30 = 2238.$$

Whence $E = 47 \cdot 3$ volts.

(3) What should be the resistance of a lamp in order that when placed on a machine of 65 volts E.M.F. $\frac{1}{6}$ of a horse-power may be expended in it?

We have from (12),—

$$R = \frac{65 \times 65}{\frac{1}{6} 746} = 33 \cdot 9 \text{ ohms.}$$

24. Relation between work, quantity, and pressure. If we have a supply of water under constant pressure, which we are using occasionally, say to drive a water-engine, we can tell how many foot-pounds of energy we have used at the end of a week, if we know the pressure and the total quantity of water used.

Similarly, if we know the electric pressure at which our electricity is supplied, and the total quantity of electricity which has passed through our circuits, we can calculate the total quantity of energy expended, or of heat produced in the resistance, however much that resistance may have been varied during the flow.

The relation is given by the equation,—

$$W = \cdot 737 \, E \, Q \quad . \quad . \quad . \quad (13)$$

Equation (13) can be derived from equation (7), when we remember that 1 ampère equals 1 coulomb per second, and 1 H.P. = 550 foot-pounds per second. The number of foot-pounds expended in sending a coulomb through the circuit, is therefore 550 times the number of H.P. expended in maintaining an ampère.

(7) thus becomes,—

$$W = \frac{550}{746} \, E \, Q = \cdot 737 \, E \, Q \quad . \quad . \quad (13)$$

Where W is the work expended or heat generated expressed in foot-pounds, E is the electro-motive force in volts, and Q is the total number of coulombs of electricity which has passed.

The equation may also be written,—

$$Q = \frac{W}{\cdot 737 \, E} \quad . \quad . \quad . \quad (14)$$

or,—

$$E = \frac{W}{\cdot 737 \, Q} \quad . \quad . \quad . \quad (15)$$

Problem (1).

With a **constant E.M.F. of 110 volts** how much work is expended in **sending 10,000** coulombs through a circuit of varying resistance?

We have from (13),—

$$W = \cdot 737 \times 110 \times 10,000 = 810,700 \text{ foot-pounds.}$$

(2) How much electricity will 33,000 foot-pounds send through a circuit with an E.M.F. of 60 volts?

We have from (14),—

$$Q = \frac{33,000}{\cdot 737 \times 60} = 746 \text{ coulombs.}$$

(3) What must be the E.M.F. in a circuit for 1474 foot-pounds to send 10 coulombs through it?

We have from (15),—

$$E = \frac{1474}{.737 \times 10} = 200 \text{ volts.}$$

Summary of formulæ. The following is a summary of the various formulæ which we have explained:—

- C stands for Current in ampères.
- E ,, E.M.F. in volts.
- R ,, Resistance in ohms.
- Q ,, Quantity in coulombs.
- H.P. ,, Rate of expenditure of work in horse-power.
- W ,, Work in foot-pounds.

1 H.P. = 33,000 foot-pounds per minute = 550 foot-pounds per second.

25. Horse-power and Work.

$$H.P. = \frac{C^2 R}{746} \qquad (4)$$

$$= \frac{E C}{746} \qquad (7)$$

$$= \frac{E^2}{746 R} \qquad (10)$$

$$W = .737 \, E \, Q \qquad (13)$$

26. Current.

$$C = \frac{E}{R} \qquad (1)$$

$$= \sqrt{\frac{746 \, H.P.}{R}}^* \qquad (5)$$

$$= \frac{746 \, H.P.}{E} \qquad (9)$$

27. Electric Pressure.

$$E = C R \qquad (2)$$

$$= \frac{746 \, H.P.}{C} \qquad (8)$$

$$= \sqrt{H.P. \, 746 \, R} \qquad (11)$$

$$= \frac{W}{.737 \, Q} \qquad (15)$$

---

* The symbol $\sqrt{}$ means "square root of" the quantity under it.

28. Resistance.

$$R = \frac{E}{C} \quad \quad \quad \quad (3)$$

$$= \frac{746 \text{ H.P.}}{C^2} \quad \quad (6)$$

$$= \frac{E^2}{746 \text{ H.P.}} \quad \quad (12)$$

29. Quantity.

$$Q = \frac{W}{737 \text{ E}} \quad \quad \quad (14)$$

30. The commercial electrical unit.—Definition. The **unit of** electrical supply is defined by the Board of Trade in the Provisional Orders to be 1000 *ampères flowing for* **one** *hour under a pressure of* **one volt**.

This is the **same as** 100 ampères under a pressure of 10 volts, **or** of **10 ampères** under a pressure of 100 volts, or generally as **1000** volt-ampères.

31. Value in horse-power per hour. This unit is mathematically **equal** to 1·34 actual horse-power working for **one hour**, i.e. **just** over 1⅓ horse-power working for one hour.

For we have by the formula (7), page 26,—

$$\text{H.P.} = \frac{E\,C}{746}$$

and where

$$E\,C = 1000$$

$$\text{H.P.} = \frac{1000}{746} = 1\cdot 34 \quad \quad \quad (16)$$

32. Value in 21-candle Swan lamps per hour. A Swan lamp as at present constructed takes exactly $\frac{1}{10}$th horse-power when working at 21 candles. Hence the commercial **unit is a** quantity of electricity that will feed **13·4** Swan lamps, **each of 21-candle power** for one hour.

Value in **14-candle Swan lamps per hour.** When lamps of smaller **candle-power are used, one** unit of electricity will feed a **proportionately larger number of** them.

One commercial **unit of electricity will feed** $\frac{21}{14} \times 13\cdot 4$ equal to 20 14-candle **Swan lamps for one hour.**

33. Equivalent in gas. Five cubic feet of gas will feed one burner of about 14 candles for one hour, 100 cubic feet of gas will feed 20 14-candle burners for one hour;
Hence
One commercial electrical unit (when feeding Swan lamps) is approximately equal in illuminating power to 100 cubic feet of gas,
Or,
Ten commercial electrical units (when feeding Swan lamps) are approximately equal in illuminating power to 1000 cubic feet of gas . . . . (27)

34. Rule for comparison of prices. We see from the above that to compare the price of electricity with that of gas we must multiply the price per electrical unit by 10, and the result will be the price of a quantity of electricity approximately equal in illuminating power to 1000 cubic feet of gas.

When, however, the electricity is used for lighting arc lamps the quantity required to produce a given light is only about $\frac{1}{10}$th of that required for incandescent lamps, and the cost in comparison with gas is proportionately reduced.

35. The Watt. The energy developed by one volt driving one ampère is called the volt-ampère or *watt*. It is equal to $\frac{1}{746}$ of a H.P., or to $\frac{1}{1000}$ of the energy which will give one unit per hour.

Questions on Chapter V.

1. What is an ampère?
2. What is a volt?
3. What is an ohm?
4. State Ohm's law?
5. How many ampères will a pressure of 120 volts send through (a) 10 ohms, (b) 180 ohms, (c) 1 ohm, (d) $\frac{1}{2}$ ohm?
6. How many ampères will 100 volts and 142 volts send respectively through the above resistances?

Questions on Chapter V.

7. A district has a total resistance of $\frac{1}{20}$th ohm, and requires 2000 ampères to light the lamps properly; what pressure is required?

8. What is the resistance of a lamp when 45 volts send 16 ampères through it? What when 120 volts send $\frac{3}{4}$ ampère through it?

9. What is the commercial unit?

10. How many units will be given by a current equal to 3 H.P. working for 2 hours?

11. How many units per hour, and how many H.P. are developed in a circuit where 1500 ampères are driven by 140 volts?

12. How many units per hour, and how many H.P. are developed in each of the cases given in Questions 5, 6, 7, and 8?

13. If Swan lamps are working at an efficiency of 210 candles per H.P., how many candles do they give per unit-hour?

14. If Swan lamps are working at an efficiency of 210 candles per H.P., how many lamps of 21 candles each can be maintained on a machine giving (a) 10 units per hour, (b) 2 units, (c) 250 units?

How many 14-candle lamps could be maintained on the same machines respectively?

15. If gas is 2s. 11d. per 1000 cubic feet, and if a gas-burner giving 14 candles burns 5 cubic feet per hour, and if Swan lamps work at an efficiency of 210 candles per H.P., what must be the cost per unit of electricity, so that the cost for the same light is the same as that of gas?

16. What is a watt?

17. How many watts are used in an arc lamp taking 45 volts and 16 ampères?

18. If an incandescent lamp uses 120 volts and ·8 ampère and gives 25 candles, how many watts per candle is it using?

19. A district uses 700 electrical H.P. at a pressure of 142 volts, what is its resistance? what current does it use?

CHAPTER VI.

SOURCES OF ELECTRICITY—BATTERIES.

The two principal sources of electric currents are (1) chemical action as exerted in batteries; (2) electro-magnetic induction as produced in dynamo-machines. In the present chapter we will speak of batteries, sometimes called *voltaic* or *galvanic* batteries, or *cells*.

36. Theory of the voltaic cell. On joining two metals either directly or by a wire, a difference of electric pressure is observed. This is a matter of observation, and I am not aware that we can say why it should be so any more than we can account for gravitation. When the metals, still joined, are partly immersed in a liquid, which acts more upon one than upon the other, the chemical action equalizes the pressures, and in doing so causes a flow of electricity along the connecting wire. The moment the equalization of the pressures has commenced, the difference is renewed again at the point or points of contact between the metals; and so, if no disturbing cause interferes, a continuous flow of electricity is kept up till the metal most acted on is entirely dissolved.

When two metals are arranged as above described in a liquid, and are in metallic communication, the one which, if alone would be most acted on, entirely protects the other, and the arrangement is called a *voltaic circuit*, or cell.

In what follows we will call that part of the metal least acted on, which is not immersed in the liquid, the positive pole of the battery, and the corresponding part of the other the negative pole.

In nearly all practical forms of the voltaic cell, the metal forming the negative pole is zinc; that forming the positive pole varies.

The simplest form of voltaic cell consists of a plate of copper and a plate of zinc (fig. 12) partially immersed in diluted * sulphuric acid, which acts on the zinc, but not on the copper. With such an arrangement, however, the current only continues for a very short time, and then ceases. Evidently some disturbing cause is acting. On examining the copper, it will be seen to be entirely coated with minute bubbles, which, if collected and tested, will be found to consist of pure hydrogen gas.

Fig. 12.

If a piece of zinc alone be dissolved in dilute sulphuric acid, the water is decomposed, and the oxygen combines with the zinc, and hydrogen is set free.

When the decomposition occurs in a voltaic cell, the hydrogen is liberated, not at the surface of the zinc, but at that of the copper.

The effect of the copper being coated with hydrogen is that a difference of pressure is no longer produced.

Why the hydrogen should appear at the copper, and why it should stop the current, is not well understood.

37. Amalgamated zinc. We have as yet assumed that all the metals used are chemically pure. The ordinary zinc of commerce of which battery plates are made is, however, not pure, but contains many particles of iron and other metals. If a piece of ordinary zinc be placed in acid, each of these pieces of iron, together with the zinc near it, forms an independent small cell, whose circuit is always closed, whether the main current is closed or open. The currents produced in these small circuits in no way help the main current, while they cause the zinc to be rapidly consumed.

---

* Unless the contrary is stated, it is to be understood that "diluted sulphuric acid" means a mixture of seven parts (by measure) of water with one part of acid (Coml. sp. gr. 1·845). In mixing, care must be taken to measure out the water first, and then to add the acid to it. It is very dangerous to add the water to the acid.

The cost of chemically pure zinc prohibits its use, so a different plan is used, which, though probably first adopted as a makeshift, is found to be in every respect equally efficacious with the employment of pure zinc.

It consists in coating the zinc with mercury. This is done by first immersing the zinc for a few minutes in dilute sulphuric or hydro-chloric acid, so as to give it a chemically clean surface, and then pouring mercury upon it. The mercury at once combines with its surface, and the whole of the zinc appears bright like silver. Zinc thus "amalgamated" is not attacked by dilute sulphuric acid, unless it forms part of a closed galvanic circuit. The precise action of the mercury is not known. It probably acts by coating the zinc and particles of iron alike with one and the same metal.

38. Constant batteries. To make a constant battery or cell, it is necessary to provide some means of freeing the positive plate from hydrogen.

39. Smee's cell (Fig. 13). In Smee's cell, which is shown in section in fig. 13, the plates consist of zinc and platinized silver, i.e. silver with a deposit of rough platinum in powder on its surface. As this presents a multitude of points, the hydrogen disengages itself more easily than from a smooth plate. As silver is much more expensive than zinc, the silver plate is usually arranged between two zinc ones, so as to use both sides of the silver, and so get a greater surface. Although the difference of pressure is independent of the size of the plates the quantity of electricity produced is not, as the larger the plates the less resistance does the battery itself offer, and therefore the more current will flow under a given pressure.

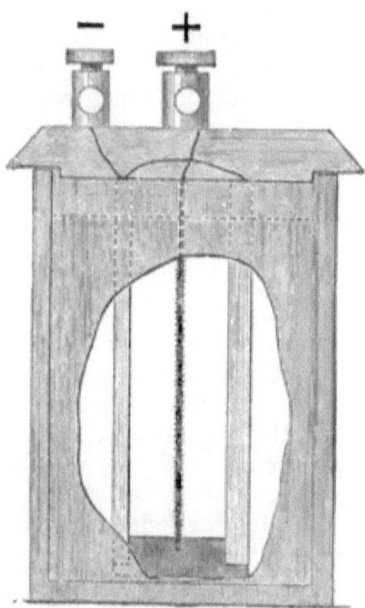

Fig. 13.

Two-fluid Cells. 37

40. Two-fluid cells. We have as yet described only cells with one fluid. In all these batteries the compounds formed by the hydrogen return to the zinc plate and retard the action upon it. Cells with two fluids are made to prevent this taking place. The two principal types are Grove's and Daniell's cells. The latter is used when a constant current of moderate strength is required for days, weeks, or months. The former, when a powerful current is required for a few hours.

41. Grove's cell. (Fig. 14.) In Grove's cell, the metals

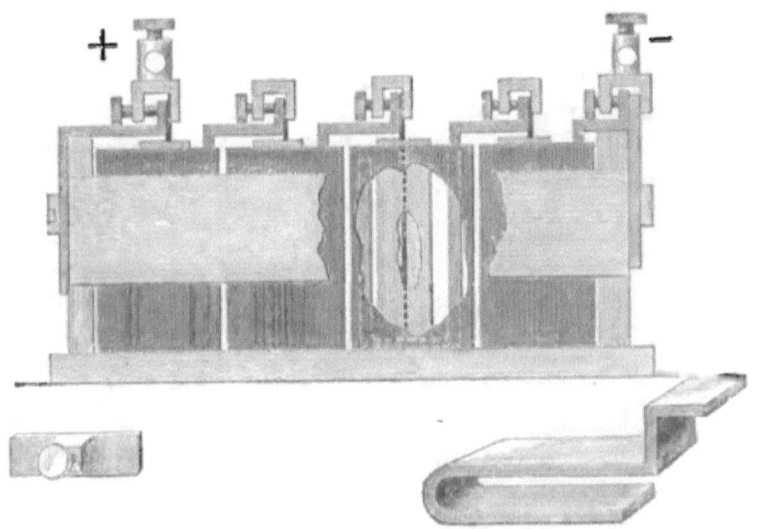

Fig. 14.

used are zinc and platinum; the fluids, strong nitric and dilute sulphuric acids. A cell of thin porous earthenware is filled with nitric acid, and contains the platinum plate. This cell is placed inside another cell, usually of ebonite, containing the zinc and dilute sulphuric acid. The porous earthenware, when wet, permits the electricity to pass freely through it, while it almost entirely prevents the liquids from mixing. In fig. 14, which shows the arrangement of the plates, several cells are represented connected together, but the reader is requested for the present to confine his attention to one only. In this cell, the hydrogen which, wherever it is set free, must be formed in the sulphuric acid, would have, in order to reach the platinum plate, to travel through the nitric acid; or even if it is only liberated on the platinum, it is still in contact with the nitric acid. The hydrogen and nitric acid at once combine, and form nitrous acid and water, both of which remain in solution in the free acid.

One of the zinc plates and one of the clamps used for holding the platinum against the zinc are shown at the bottom of fig. 14.

Grove's battery has been hitherto the only voltaic arrangement used for purposes where great power is required.

42. Daniell's cell. (Fig. 15.) In this cell the metals used are zinc and copper. The former is usually immersed in dilute sulphuric acid; the latter in a saturated solution of sulphate of copper. In a very convenient form of the cell shown in fig. 15, the zinc in the form of a rod is placed inside the porous cell, and the containing vessel being made of copper acts as the other plate.

Inside the copper cell and near the top is a copper shelf, perforated with many holes. This shelf serves to keep the porous cell in its place. On it are piled up a number of crystals of sulphate of copper. The cell is filled with a saturated solution of the same, i.e. with water in which is dissolved the maximum quantity of sulphate of copper which it will contain.

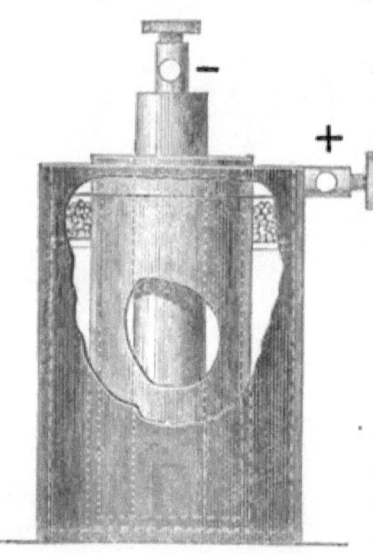

Fig. 15.

In the inner cell is the zinc rod, and, according to the purpose for which the battery is required, either dilute sulphuric acid, salt and water, or plain water; the latter, while causing a great diminution of power, increasing the constancy of the battery.

In the cell from which fig. 15 is drawn, the copper cylinder is 6 inches deep and 3 inches in diameter.

When the circuit is closed, the hydrogen, whether it comes from the zinc through the porous cell towards the copper, or is liberated on the copper, meets the solution of sulphate of copper, and, taking from it an atom of sulphur and four atoms of oxygen, forms sulphuric acid, and liberates metallic copper, which is deposited on the copper plate. At the same time sulphate of zinc is formed in the sulphuric acid cell.

Thus the thickness of the zinc plate diminishes, and that of the copper plate increases, while the cell is worked.

Now for each grain of copper deposited on the plate, a corresponding quantity of sulphate of copper is destroyed, and the solution gets weaker. As soon as this occurs, a portion of the sulphate of copper on the shelf is dissolved. The part of the liquid which has dissolved it becomes denser than the rest, and sinks to the bottom, and a fresh portion of the weakened solution comes in contact with the solid sulphate of copper, and so a circulation goes on till all the liquid is again saturated.

The power of this cell steadily diminishes until the dilute acid is saturated with sulphate of zinc, after which it remains almost constant for a very long time. For this reason, when constancy is more important than strength, it is customary to saturate the solution with sulphate of zinc before beginning work.

Numerous other forms of Daniell's cell are in use, the various modifications having been introduced with a view of preventing the mixing which goes on through the walls of any porous cell, and because of the resistance which such a cell offers to the electrical and chemical action.

43. Latimer Clark's standard cell. Mr. Latimer Clark has devised a "Standard Cell,"[*] that is, a cell

---

[*] Phil. Trans., 1874, p. 1.

whose electric pressure is always constant. Great difficulty had been experienced in determining a practical unit of electric pressure, owing to the fact that not only are there differences in the electric pressures of different ordinary cells, supposed to be of the same construction, but that the electric pressure of the same cell varies from day to day.

With the "standard cell" it is found that as long as it is not used to produce a current, the electric pressure between its poles remains absolutely constant. The maximum difference observed in a series of comparisons between different models of it during a period of several months was not more than $\frac{1}{1000}$ part of the whole electric pressure, and it appears that even this difference might be accounted for by an accidental difference of temperature.

The electric pressure which the cell gives at 15°·5 C. is taken as the standard.*

It is found that the force decreases with increase of temperature, and that the rate of variation for 10° above and below 15°·5 is 0·6 per cent. for each degree centigrade.

"The battery is formed by employing pure mercury as the negative element, the mercury being covered by a paste made by boiling mercurous sulphate in a thoroughly saturated solution of zinc sulphate, the positive element consisting of pure distilled zinc resting on the paste.

"The best method of forming this element is to dissolve pure zinc sulphate to saturation in boiling distilled water. When cool, the solution is poured off from the crystals and mixed to a thick paste with pure mercurous sulphate, which is again boiled to drive off any air; this paste is then poured on to the surface of the mercury, previously heated in a suitable glass cell; a piece of pure zinc is then suspended in the paste, and the vessel may be advantageously sealed up with melted paraffin wax. Contact with the mercury may be made by means of a platinum wire passing down a glass tube, cemented to the inside of the cell, and dipping below the surface of the mercury, or more conveniently by

---

* It equals 1·42 (new) volt. See page 22.

a small external glass tube blown on to the cell, and opening into it close to the bottom. The mercurous sulphate (Hg_2SO_4) can be obtained commercially; but it may be prepared by dissolving pure mercury in excess in hot sulphuric acid at a temperature below the boiling-point: the salt, which is a nearly insoluble white powder, should be well washed in distilled water, and care should be taken to obtain it free from the mercuric sulphate (persulphate), the presence of which may be known by the mixture turning yellowish on the addition of water. The careful washing of the salt is a matter of essential importance, as the presence of any free acid, or of persulphate, produces a considerable change in the electro-motive force of the cell."

44. Batteries of several cells. We have said that when the circuit is open (that is, when the poles are not connected), the pressures of the poles of any cell differ by a quantity which is approximately constant for each kind of cell. We often, however, require a difference of pressure greater than can be given by any one cell. This is obtained by connecting a number of similar cells "in series," that is, connecting the positive pole of one with the negative pole of the next, and so on—a number of cells so connected is called a voltaic *battery*. Fig. 14 is a representation of a Grove's battery of four cells. It is seen that the zinc of each cell projects sideways over the next, and the platinum of that cell is clamped to it. The only reason why the zinc plates are chosen to project rather than the platinum, is the far greater expense of the latter, and the fact that owing to their not being consumed, it is only necessary to make them of the thickness of writing-paper, when of course they have but small rigidity.

Thus all the poles are connected two and two, except one from each of the end cells. These two free poles are called "the poles of the battery."

Their difference of pressure is as many times the difference of pressure between the poles of a single cell, as there are cells in the battery, i.e. in a battery of 4 cells, if we suppose the difference of pressure between two poles of the same cell to be represented by 2 volts, that between the poles of the battery will be represented by 8 volts; if there are 5 cells, by 10 volts, and so on.

45. Conventional sign. To save repeating pictures of the battery, the conventional sign (fig. 16) is used in the diagrams; the thin and thick lines representing respectively the zinc and other plates, and the number of them the number of cells.

Fig. 16.

Questions on Chapter VI.

1. Describe a simple voltaic cell.
2. How is zinc amalgamated?
3. What is required to make a battery "constant"?
4. Describe Smee's cell.
5. Describe Grove's cell.
6. Describe Daniell's cell.
7. For what purpose is Latimer Clark's standard cell used?
8. What symbol is commonly used in diagrams to denote a battery?

CHAPTER VII.

SOURCES OF ELECTRICITY—ELECTRO-MAGNETIC INDUCTION.

It will be remembered that in Chapter III. we have spoken of the lines of force which emanate from the poles of magnets, and have shown how their directions can be traced, and we have stated that the space surrounding a magnet, and through which the magnetic lines radiate, is called a *magnetic field*.

46. Electro-magnetic induction. If a wire be moved through a magnetic field, an electric pressure will be produced between its ends which will be simply proportional to the number of lines of force cut by it per second; lines cut in one direction being reckoned +, those cut in the other direction being reckoned −, the sign being also reversed if the polarity of the field is reversed.

The number of lines of force cut per second depends,

First (in a uniform field), on the length measured in a straight line from one end of the moving wire to the other; i.e. if A B C., fig. 17, be the wire, it depends on the length A D C.

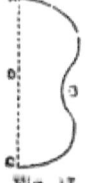

Fig. 17.

Second, on the angle which the direction of motion makes with the lines of force. If the wire moves along the lines of force, it will cut none of them; if at right angles to them, it will cut the maximum number. The number cut is directly proportional to the sine of the angle which the direction of motion makes with the lines of force.

Third, on the number of lines of force which pass through each unit of area of the region across which the wire moves, i.e. on the strength of the magnetic field.

Fourth, on the velocity of motion.

If the ends of the wire are connected by another wire not in motion, a current will flow through the wire, whose strength depends on the electric pressure produced and on the resistance of the circuit.

For instance, the circuit may be completed by means of two fixed rails, on which the moving wire slides, the rails being connected at one end.

If in a uniform field the ends of the wire are connected by means of a second wire, which also moves across the lines of force (fig. 18), no current will be produced. The electric pressures in the two halves of the ring formed by the two wires will be in the same absolute direction in space, and therefore in opposite directions in the ring. In fig. 18 we suppose the lines of force to be perpendicular to the paper, and to be represented by the dots. We suppose the circuit to consist of a curved wire whose ends are connected by a zigzag one, and that it moves in the direction of the large arrow. We see that each half of the circuit cuts the same number of lines of force, and the electric pressures are both of the same magnitude and are in opposite directions in the ring, as represented by the small arrows; and therefore there will be no current.

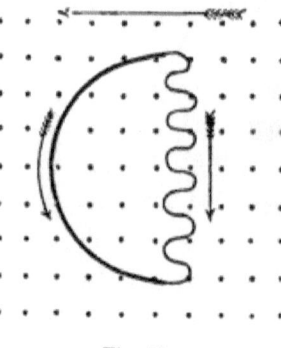

Fig. 18.

The moving of a wire across the lines of terrestrial magnetic force will produce an electric pressure between its ends. A large portion of the earth's magnetic force is vertical; a horizontal wire moved horizontally and parallel to itself will therefore cut terrestrial magnetic lines. Let us suppose the rails of a railway to be insulated from each other, but connected at one end through a galvanometer. The wheels and axle of a railway carriage would complete the circuit. As the carriage moves, an electric pressure will be produced between the ends of the axle, which will produce a current **through the** galvanometer. If there are several axles they **will all act** in the same direction (like batteries in parallel circuit).

If, instead of being connected **to** the rails, the galvanometer **were** connected to the ends of the axle, and carried in the carriage, no current would be produced, the reason being that equal electric pressures would be produced **in** the axle and in the connecting wires of the galvanometer. These pressures would be in the same absolute direction, and therefore would be opposed **to** each other in the circuit.

47. Theory of **electric generators.** All electric generators consist of machines for moving wires past **magnets,** or magnets past wires, **the** connections being **so** arranged that the electro-motive forces generated may **produce currents.**

If **our moving circuit consists** of a ring which is sufficiently **large in comparison with** the field, we can cause one **side** of it to move over the **N. pole** of a magnet while the other side is moving over the **S. pole,** and the electric pressures produced in **the two halves will** then be in opposite directions in **space, and therefore in the same** directions in the ring, **and currents will circulate.**

Let us suppose our moving circuit to consist of the two axles of **a four-wheeled railway carriage** (fig. 19), connected

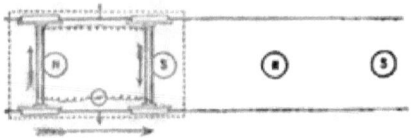

Fig 19.

by wires running along the sides of the carriage, and passing through **a** galvanometer carried in it ; **and** suppose that instead **of** making use **of** terrestrial **magnetism, we** bury in the permanent **way a** number **of** powerful magnets of alternate polarity, **the** distance between the poles being **equal to** the distance between the axles. We see that the electric pressures produced in the two axles will be in opposite directions, and therefore a current will circulate until the axles arrive at the neutral point between two poles. The current will then diminish to nothing, and then gradually increase again; but as the field now being passed through by each axle is of opposite polarity to what it was before, the current will be in the opposite direction. And thus as the carriage moves on, currents will be produced which will be reversed in direction each time the centre of the carriage passes a pole.

The principle of this arrangement is the basis of all *alternating-current machines.*

We see, therefore, that the only way in which currents can be induced in a closed ring or coil of wire, is by the approach to or recession of the coil from a pole.

A motion through a uniform field produces equal and opposite electric pressures in the two sides of the coil or ring, which pressures neutralize each other.

If the floor of the carriage in fig. 19 had consisted of a thick iron plate, the electric pressure produced would have been greater, for the lines of force would have been

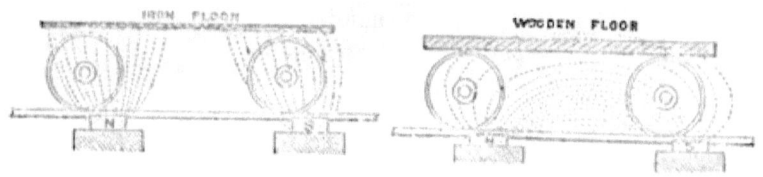

Fig. 20.   Fig. 21.

nearly vertical, as in fig. 20, instead of partly horizontal, as in fig. 21.

In practice, the only way in which wires can be moved rapidly past magnets is by attaching them to the rim of a revolving wheel round which stationary magnets are arranged, so that the wires pass the same magnets again and again.*

* Or the magnets may be attached to the revolving wheel and moved past fixed wires.

Another way of enabling the induced electro-motive forces to produce currents, is that invented by Professor Paccinotti in 1863, and known as the "Paccinotti or Gramme ring." By this apparatus the currents are produced continuously in one direction. Its principle is the basis of all *continuous-current machines*. We shall fully describe it in Chapter XVIII.

48. Direction of the induced currents.—Lenz's law. In 1834 Lenz enunciated the following remarkable law:—

Whenever a current is induced in a circuit by the relative **motion of** *the circuit and* **of** *a magnet,* **or** *of another circuit carrying* **a current,** *the direction of the induced current is such that by its attraction* **or** *repulsion on the inducing* **magnet or** *circuit it opposes the motion.*

We see that if this **were not so** we should have a "perpetual motion," **as** the **induced** current might produce the motion which itself produced the induced current.

49. Induction by variation of current in one of two stationary circuits. If an electro-magnet, or a circuit (which may be regarded as an electro-magnet without an iron core) be placed **near** a coil **of wire,** and the current in the electro-magnet be made to vary, currents will **be** induced in the circuit as long as the variation continues.

The direction of the current produced **by** increasing magnetism **is the same as that** produced by **an** approaching pole, that of **the** current **produced by decreasing** magnetism is the same **as that** produced by a receding **pole.** An increasing current in the electro-magnet induces a current in **a** direction opposite to its own; a decreasing current, one in **the same** direction. The action **of** the *induction coil* * depends **on** this law.

50. Iron core. An iron core may be placed in **the** circuit in **which a current is to be** produced. This generally increases **the** effect as **it** strengthens and concentrates the lines of force.

51. Effect of the iron core on the coil surrounding it. If the magnetism **of** the **iron core** is altered, as, for instance, **by** moving a magnet **to and** from it, currents will flow in the coil as long as the change continues.

* See Chapter XXI.

While the magnetism of the core is *increasing*, the direction of the induced current will be such that it will tend to make the iron core a magnet having *opposite* polarity to that actually caused by the induction of the neighbouring magnet.

While the magnetism is *decreasing*, the direction of the induced current will be such that it will tend to make the iron core a magnet having the *same* polarity as that actually caused by the induction of the neighbouring magnet.

52. Dynamo machines. Generators of electricity based on the principle of electro-magnetic induction are called *dynamo machines*. They all consist of machines for moving magnets past coils, or coils past magnets, with considerable velocity, as we have already stated. The only practical way of carrying out such motion is to make it circular, and, generally speaking, all dynamos consist of two circular rings or wheels, on one of which electro-magnets are fixed, and on the other coils of wire. One of these wheels is fixed, and the other is made to revolve by suitable machinery.

Dynamos are of all sizes, from toys which can be turned by one hand to machines which will work 5000 lamps each, and each take a 500 H.P. steam-engine to drive it.

Some details of various dynamos will be given in a later chapter.

Questions on Chapter VII.

1. What happens if a straight wire is moved through a magnetic field?

2. What happens if a ring of wire be moved in its own plane through a uniform magnetic field?

3. In (1) what happens if (a) the speed is increased, (b) the strength of the field is increased?

4. State Lenz's law.

5. Show that if Lenz's law did not hold, we could obtain "perpetual motion".

6. A coil of wire forms a closed circuit, what happens when one end of a magnet is (*a*) suddenly pushed into it, (*b*) suddenly withdrawn.

7. A coil of wire has an iron core; what is the effect of varying the magnetism of the core?

8. Let a closed coil of wire with an iron core be set up on end, and a watch laid on its back near it, will the direction of the induced currents be the same or opposite to that of the motion of the hands of the watch when a steel magnet is moved as follows with respect to the **upper** pole of the core,—

      (*a*) N. pole lowered towards it,
      (*b*) S. ,, ,, ,, ,,
      (*c*) N. pole raised from it,
      (*d*) S. ,, ,, ,, ,, ?

CHAPTER VIII.

MEASUREMENT OF ELECTRIC CURRENTS—GALVANOMETERS.

53. Galvanometers. We have spoken of the different strengths of the currents produced by various batteries; we will now give an account of some of the methods used to measure them.

The instrument most commonly used to measure currents is called a "galvanometer."

Fig. 22.

Galvanometers may be roughly divided into two classes—

(1) Those used to measure strong currents.

(2) Those used to detect feeble currents.

The latter class are of great importance, for we shall see that most measurements of resistance are determined by a balancing of currents, such that, when the equilibrium is complete, no current shall pass; and, therefore, it is important to detect even the feeblest current.

54. The tangent galvanometer. Among the first class the tangent galvanometer stands preeminent.

In its simplest form it consists of a large vertical ring of copper wire (fig. 22), in the centre of which is a small compass needle. When the instrument is used, it is turned so that the ring lies in the magnetic meridian, and therefore the needle lies in the plane of the ring. When an electric current is sent round the ring, it deflects the needle.

Mathematical reasoning * shows us that the strength of the current is proportional to—

(1) A quantity depending on the construction of the galvanometer, and called its "constant." This quantity can be determined once for all for each instrument, and marked on it.

(2) On the strength of the earth's magnetism for that time and place of observation. This quantity can be obtained from the constant records taken at Kew Observatory. Its variations, however, are so small, that for most observations it can be considered as a constant quantity.

(3) To the *tangent* of the angle of deflection. The nature of the tangent of an angle is fully explained in all books on trigonometry, so need not be explained here. Tables are published giving the tangent of every angle, and it is necessary to have such a table to refer to in measuring currents by the tangent galvanometer.

The single ring is the simplest form of the galvanometer, but it is not suited to accurate work, owing partly to the want of security as to the ring remaining perfectly plane and rigid, and partly to the comparatively high ratio which the irregular action of the connecting wires bears to the regular action of the ring.

The following is a better form of the instrument (fig. 23):—

Two rings are used, one on each side of the needle, so placed that the centre of the needle is at the centre of their common axis (or line joining their centres), and that this line, when the instrument is adjusted, is at right angles to the magnetic meridian, i.e. lies magnetic E. and W. The rings are made of wood, and the wires are wound on them. In a groove at the outside is a massive copper ring, which is used only for rough experiments with very powerful currents. The rings on each side are connected so that the current goes in the same direction

* See my "Electricity," 2nd edit. vol. i, p. 245.

through each. The inside of each ring is turned so as to make part of a cone, such that, if it were continued, its

Fig. 23.

apex would be at the centre of the needle. On this three coils of wire, of respectively 3, 9, and 27 windings, are laid; they are, of course, covered with silk or cotton to make the current go round and round, instead of merely across from one wire to another. Each winding produces its own effect on the magnet, and thus with a current, such that its effect when in the single ring is unity, we can produce an effect on the needle equal to 3, 9, or 27, or any combination of the sums or differences of those numbers.

The needle is a short pointed one with a piece of agate let in to the top of the cap where it rests on the pin, while to allow a larger divided circle to be used, a light aluminium needle is attached to it.

Galvanometers.

The needle is arranged so that the points of the aluminium needle are as nearly as possible in the line of magnetization. Any error in this adjustment is, however, corrected by taking alternate readings with the current in opposite directions; the one reading will then be as much too great as the other is too small, and the mean will be the true reading.

Let us take an extreme case as an illustration, and suppose that the angle between the magnetic axis of the needle and the line joining the points is 1°, and let us suppose that the current causes a deflection of 30° of the magnetic axis (fig. 24). Then the reading in one direction will be 29°, and in the other 31°. The mean is 30°, the true deflection; and we see that to determine this does not require a knowledge of the angle between the direction of the pointer and the magnetic axis of the needle. This can, however, be easily obtained if wanted, for it is evidently half the difference between the two readings.

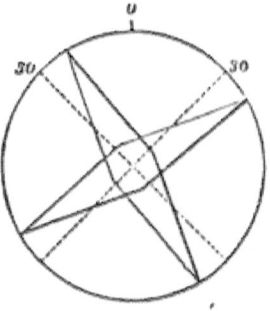

Fig. 24

The instrument is supported on three levelling screws, and the base and supporting pillar are connected by a pivot, which enables the latter to be turned round so as to adjust the circles in the meridian.

55. Sensitive galvanometers.—Astatic needles.

Under the head of *sensitive galvanometers* comes all those used for detecting or measuring feeble currents. The *astatic needle* is an arrangement used in most galvanometers of this sort to diminish the earth's action, while at the same time it increases the action of the current. The needle consists of two magnets, almost, but not exactly, of the same strength, connected together by a rigid bar, with their similar poles in opposite directions. They are usually not pivoted, but hung by a silk thread. The marked end of the stronger magnet will point to the north, but if the combination be deflected by any means, the force tending to bring it back to the meridian will only be the difference of the forces exerted by the earth on the two magnets respectively. The coil of wire through which the current passes has an opening left near the centre of the top side, and the connecting bar of the magnets passes through it. One magnet thus hangs inside the coil, and the other just above it (fig. 25). A reference to fig. 6 will show that the actions of the top and bottom of the coil on the lower needle are in the same direction, while, though the actions on the upper needle are in opposite directions, that of the top of the coil which is nearest, and therefore most powerful, is in the same direction as those on the lower one.

In sensitive galvanometers the current goes many, often several thousand, times round the needle.

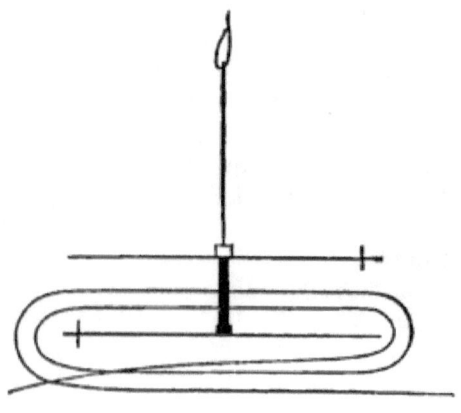

Fig. 25.

56. Sir Wm. Thomson's reflecting galvanometers. —The lamp, scale, and mirror. To detect and measure small angular deflections of a needle, a long pointer is necessary; but, if a long material pointer were attached to the needle, its weight would destroy the sensitiveness of the instrument.

Sir Wm. Thomson has therefore arranged a method by which a *beam of light* is made to act as a pointer of any length, and absolutely without weight.

A circular mirror, about ⅕ of an inch in diameter, is rigidly attached to the needle, or rather the needles are cemented on to the back of the mirror.

A lamp and scale, of which the back (that is, the side furthest from the galvanometer) is shown in fig. 26, is placed on the table about two feet from the instrument. The light passes through a small opening in the lower part of the scale, falls on the mirror, and is reflected on to the upper part, making a spot of light. The least motion of the needle and mirror, of course, moves the spot along the scale. The distance which it moves is equal to that which would have been traversed by the end of a pointer whose radius was double the distance from the mirror to the scale.*

Fig. 26.

The aperture through which the light passes is sometimes a vertical slit, sometimes a round hole, with or without a vertical wire stretched across it.

Sometimes the mirror is plane, and the light is brought to a focus on the scale by means of a lens. Sometimes the mirror is concave, and the lens is dispensed with.

When the slit is used, the moving image is a vertical line of light; when the hole is used, it is a bright disc crossed by a fine vertical black line, the image of the wire.

The scale is usually divided into millims., and printed black on white glazed paper.

In using a flat-wicked paraffin-lamp, the wick should be placed "edgeways"—that is, at right angles to the scale.

These galvanometers are sometimes made astatic, some-

---

* Any boy who has ever amused himself by throwing sunlight from a looking-glass into his neighbour's windows will know what a small motion of the looking-glass is sufficient to give a large motion to the spot of light.

Fig. 27.

times not. The mirror is usually less than ¼ in. in diameter, and is very thin. In the non-astatic form (fig. 27) the magnet, or rather magnets, for several are generally used, are cemented to the back of the mirror, and are usually about ⅛ in. in length. The whole system of magnets and mirror weighs less than a grain. The object of having several magnets is to get the maximum of magnetization*with the minimum of weight. The mirror and magnets are hung by a single fibre of unspun silk in the centre of a circular coil of wire, which is enclosed in a brass cylinder. The front end of the cylinder is closed by a glass plate, the back by a brass one, in the centre of which is a small disc of plate glass, through which the mirror can be seen. The cylinder is supported horizontally on a tripod stand, each leg of which is furnished with a levelling screw, by which the apparatus is adjusted until the mirror is seen to swing clear.

To avoid the inconvenience of having to place the apparatus always in the magnetic meridian, a large curved magnet feebly magnetized is supported horizontally on a vertical stem, fixed to the top of the case.

The magnet can be turned by hand on the bar on which it slides somewhat stiffly; and by its directive force makes an artificial magnetic meridian in any desired direction. A fine adjustment is obtained by moving the stem itself by means of a tangent screw. The magnet can also be slid up or down the stem so as to act more or less powerfully on the suspended magnet.

This galvanometer is made sometimes with a short thick wire, sometimes with a long thin one.

In the astatic form (fig. 28), which is used only with fine wire galvanometers, each needle is surrounded by its own coil of wire. The current of course goes opposite

---

* This is required because the more highly the needle is magnetized the more rapidly will it come to rest after being set swinging.

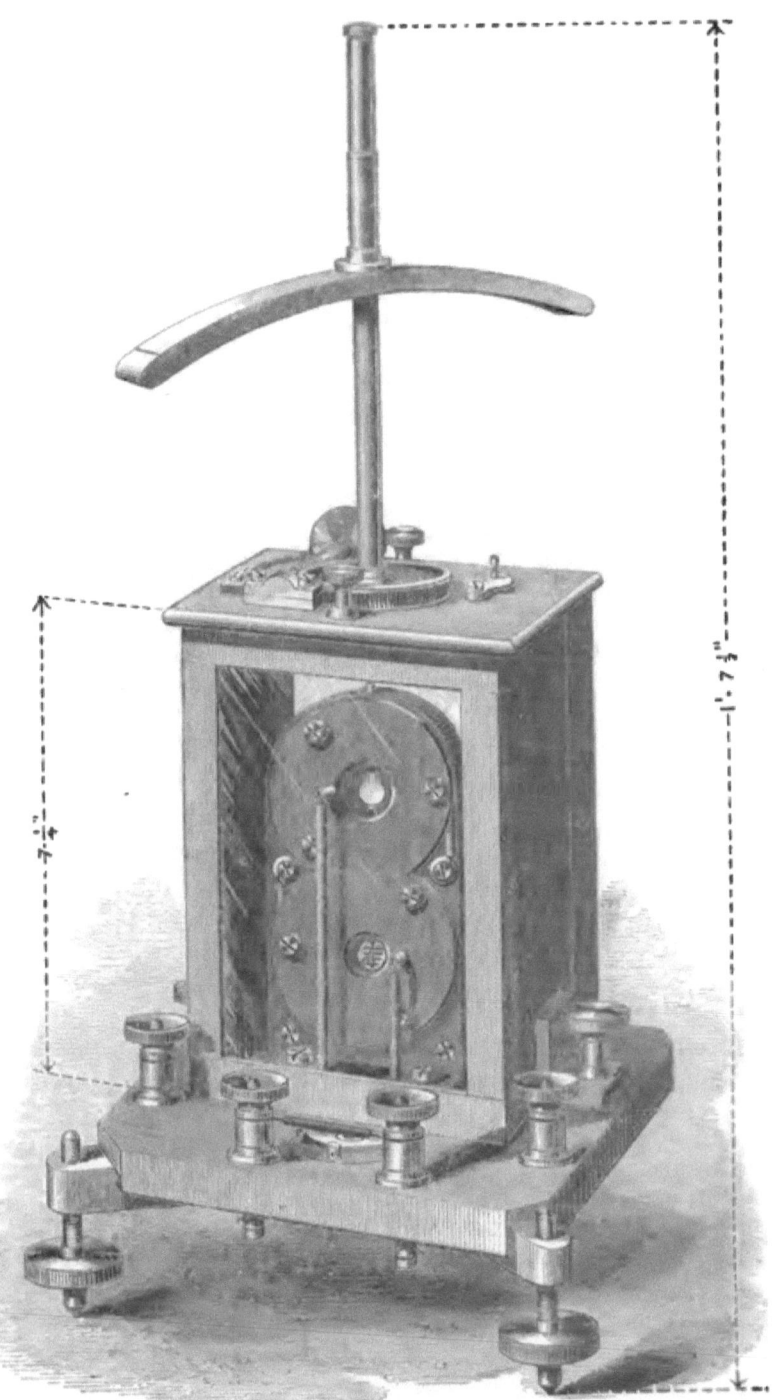

Fig. 28.

ways in these two coils. The coils are sometimes enclosed in a vertical cylinder of glass, and sometimes in a square glass case. As the magnet and mirror system is necessarily somewhat heavier in this construction, an aluminium fan is sometimes attached to it to check its vibrations. The other details are similar to those of the tripod form.

57. Measurement of strong currents.—Siemens' electro-dynamometer. Numerous instruments are in use for measuring strong currents, but we will only here describe one, namely, the Siemens *electro-dynamometer*.

The principle on which the electro-dynamometer is founded is the fact that two neighbouring wires carrying currents attract each other if the currents are in the same direction, and repel if they are in opposite directions.

The instrument as constructed by Messrs. Siemens consists of a fixed coil of wire (fig. 29) of the shape of a flattened ring, and a ring of one or more turns of stout wire suspended by a thread and a spiral spring. The plane of the suspended ring in its position of rest is at right angles to that of the fixed ring. The two ends of the suspended ring dip into mercury cups, which allow a current to be sent round it while it is still quite free to turn. The wires are connected so that a current entering the instrument passes through both the fixed and suspended coils.

The ring suspended by the spiral spring has its upper end attached to a nut or button called a "torsion head." The latter carries a pointer, which, when the torsion head is turned by hand, moves over a scale of degrees, and indicates through what angle the top end of the spring has been twisted.

When a current is sent through the instrument the suspended coil is deflected, but is prevented moving more than about 5° by a stop. The torsion head is then turned by hand until the twist or torsion of the spring, acting against the current, brings the suspended ring back to its zero position. The number of degrees through which the torsion head has had to be turned is a measure of the strength of the current. A table is supplied with each instrument, showing the number of ampères corresponding to each degree of twist. The table is prepared by comparing the indications

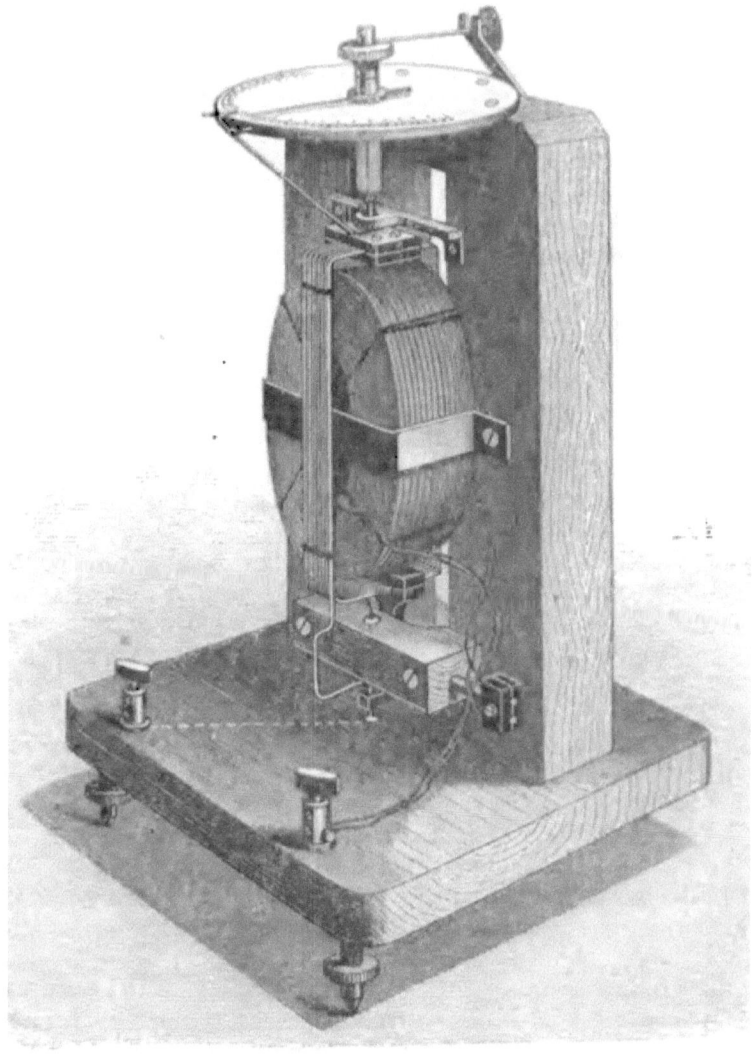

Fig. 21.

of each instrument with those of an absolute electro-dynamometer,* when the same currents are sent through both instruments. Some of the instruments have two fixed coils, one consisting of a good many turns for feeble currents, the other of a few turns for strong currents. Such instruments of course have two reduction tables.

Thus to measure a current with this instrument, we first level the instrument carefully, and adjust it so that the suspended coil hangs at its zero position. If the instrument is in proper order, this will be when the torsion pointer is also at zero. We then send the current through it, and then turn the torsion head until the suspended coil returns to zero. We then look in the table to see what current corresponds to the reading of the torsion pointer.

It sometimes happens that owing to the instrument being a little out of order, the torsion head has to be turned a few degrees from zero, in order to bring the suspended coil to its zero when no current is passing. When this has to be done, the zero error must be subtracted from or added to the reading of the torsion needle, to give the amount of torsion balancing the current.

For instance, suppose when no current is passing, that in order to bring the coil to zero, the torsion needle has to be moved $4°$ in the same direction as that in which it is afterwards to be moved to balance the current; and that its position when the current is balanced is at the $20°$ mark.

Then, in order to balance the current, we have moved the torsion needle from $4°$ to $20°$, that is through $16°$, and our current will be that corresponding not to $20°$, but to $16°$ in the table.

If we had previously had to move the torsion head $4°$ in the opposite direction and it balanced the current at $20°$, we should have had to move it from $-4°$ to $20°$, i.e. through $24°$; and our current will be that corresponding to $24°$ in the table.

The chief advantage of the instrument is that it measures "alternating" currents as well as direct ones, for the attraction simply depends on the currents in the coils being in

* See my "Electricity," 2nd edit. vol. ii. p. 79.

Siemens' Electro-dynamometer.

ing the number of ampères corresponding to each degree of twist. The table is prepared by comparing the indications

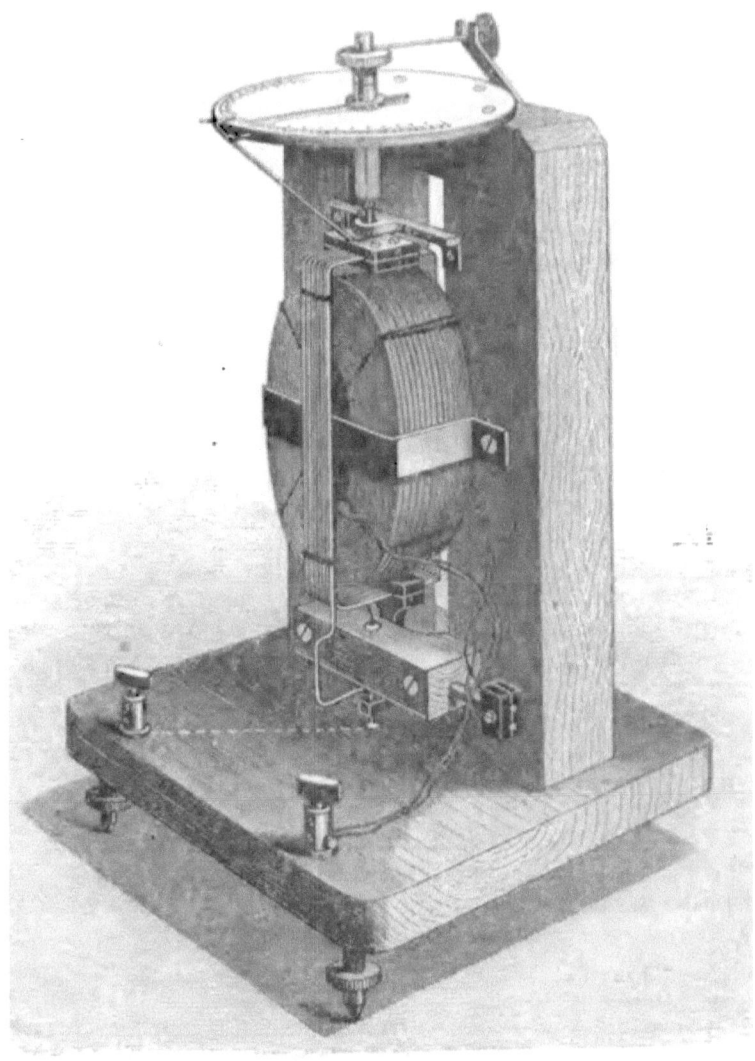

Fig. 21.

of each instrument with those of an absolute electro-dynamometer,* when the same currents are sent through both instruments. Some of the instruments have two fixed coils, one consisting of a good many turns for feeble currents, the other of a few turns for strong currents. Such instruments of course have two reduction tables.

Thus to measure a current with this instrument, we first level the instrument carefully, and adjust it so that the suspended coil hangs at its zero position. If the instrument is in proper order, this will be when the torsion pointer is also at zero. We then send the current through it, and then turn the torsion head until the suspended coil returns to zero. We then look in the table to see what current corresponds to the reading of the torsion pointer.

It sometimes happens that owing to the instrument being a little out of order, the torsion head has to be turned a few degrees from zero, in order to bring the suspended coil to its zero when no current is passing. When this has to be done, the zero error must be subtracted from or added to the reading of the torsion needle, to give the amount of torsion balancing the current.

For instance, suppose when no current is passing, that in order to bring the coil to zero, the torsion needle has to be moved 4° in the same direction as that in which it is afterwards to be moved to balance the current; and that its position when the current is balanced is at the 20° mark.

Then, in order to balance the current, we have moved the torsion needle from 4° to 20°, that is through 16°, and our current will be that corresponding not to 20°, but to 16° in the table.

If we had previously had to move the torsion head 4° in the opposite direction and it balanced the current at 20°, we should have had to move it from −4° to 20°, i.e. through 24°; and our current will be that corresponding to 24° in the table.

The chief advantage of the instrument is that it measures "alternating" currents as well as direct ones, for the attraction simply depends on the currents in the coils being in

* See my "Electricity," 2nd edit. vol. ii. p. 79.

the same **direction, and** is not affected if they are both reversed. **This** is important, as a large class of the machines used in electric lighting give currents whose direction is reversed **many** times a second.

Questions on Chapter VIII.

1. Describe the tangent galvanometer.
2. What is an astatic galvanometer.
3. Describe the reflecting galvanometer?
4. Describe **the Siemens** electro-dynamometer.

CHAPTER IX.

MEASUREMENT OF ELECTRIC PRESSURE OR ELECTRO-MOTIVE FORCE.

58. Electric pressure so closely resembles water pressure that we shall more easily understand the methods used to measure it if we first discuss methods by which water pressure, or the difference of pressure between two points in a stream of water, might be measured.*

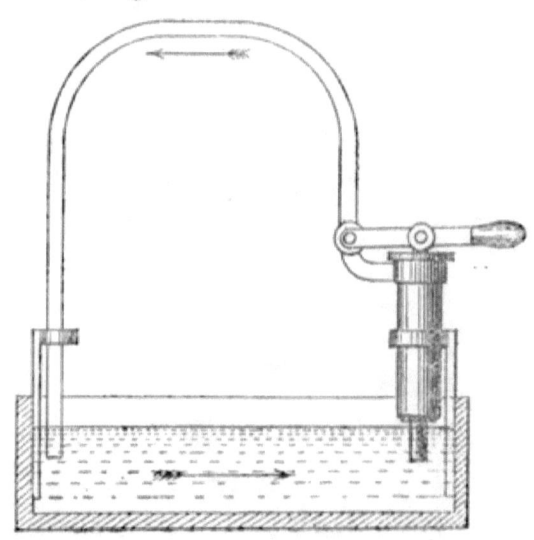

Fig. 30.

* It must be understood that the conditions of water and electric pressure are not precisely similar, and that therefore these three

The first is what may be called the **Condenser or reservoir method.***

Suppose we have a current of water being forced round a pipe by a pump, as in fig. 30, we see that the quantity of water which will circulate per minute will depend on the pressure applied by the pump. Now suppose we cut the pipe at the top, as in fig. 31, and insert in the cut part a cylinder with an elastic membrane (M), such as a piece of sheet india-rubber, stretched across it.

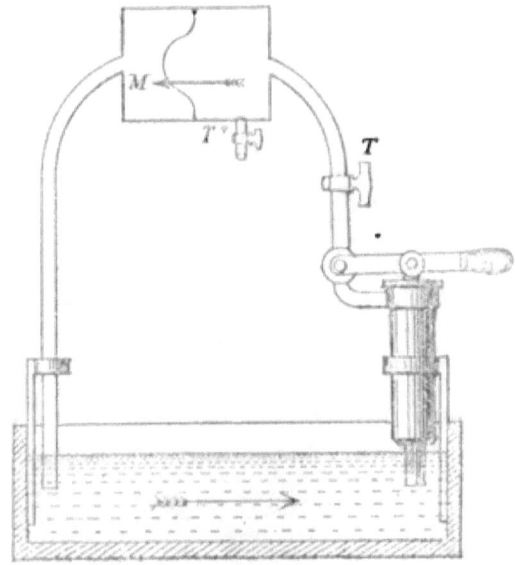

Fig. 31.

The water can no longer circulate, but the pumping may be continued for a short time, until the membrane is as much stretched in the direction of the arrow as the particu-

methods have been selected, not as the three best methods of measuring water pressure, but as those which most closely resemble the three best methods of measuring electric pressure.

* To prevent possible misapprehension, it may be here stated that a condenser is quite a different thing to a "storage battery" or an "accumulator."

lar pressure then being applied by the pump is able to stretch it. We see that the greater the pressure the more the membrane will be stretched, and the more water will enter the right-hand side of the cylinder.

If, when the stretching has proceeded as far as the pump is able to carry it, we close the tap. T_1 the apparatus will remain "charged" and with the membrane stretched for an indefinite time.

If now we open the tap T_2 the elasticity of the membrane will cause it to straighten itself, and the quantity of water which flows out or is *discharged* will depend on how much the membrane has been stretched, or, in other words, *the quantity discharged will measure the pressure which has been employed in charging.*

59. Condenser or reservoir method of measuring electric pressure. We note that the cylinder in fig. 31 forms two reservoirs for water which are separated by a membrane which can be stretched by water-pressure, allowing more water to enter one half of the cylinder, the excess of water in one half depending on the difference of pressure on the two sides.

In the electrical case the cylinder is represented by two conductors of electricity, such as two sheets of tinfoil, separated by something which the electricity cannot pass through, but which can be electrically stretched according to the electric pressure applied to it, that is, according to the excess of electric pressure applied to one tinfoil over that applied to the other.

For this something, glass, mica, and paraffined paper are used.

Fig. 32 shows how electric pressure may be measured by such an apparatus.

We note that as the ends of the water-pipes dip into an open tank, so the ends of the electric wires are connected to the earth.

B is the battery which generates the pressure we wish to measure.

C is a piece of glass with a sheet of tinfoil pasted on each side of it, called a *Condenser*.

The electric pressure from B "charges the condenser," that is, electrically **stretches** the glass. By means of the

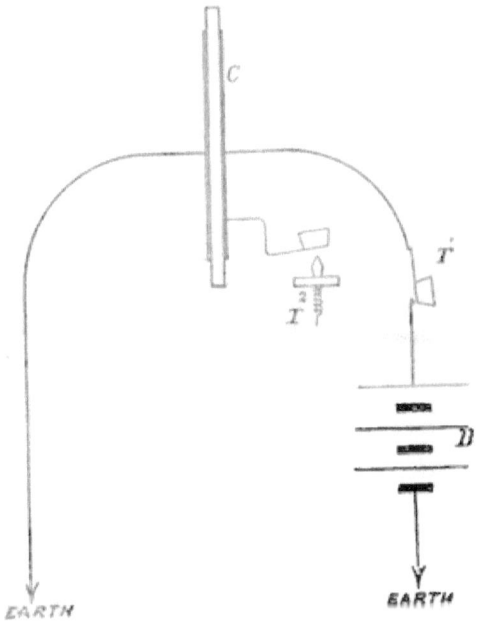

Fig. 32.

tap T_1, **the** condenser **is** separated or **insulated** from the battery, and by means **of** a tap T_2, the condenser is then "discharged" through an apparatus which measures the quantity of electricity sent through it. This quantity is **a** measure of **the** electrical stretching of the glass, and is proportional to **the pressure by** which the condenser was charged.

60. Practical form of the above experiment. The following is **the manner in** which the experiment above described is **carried** out in practice.

For accurate work the condenser requires **to** have a large surface. Instead of one large sheet it is, for convenience, constructed of a large number **of** alternate sheets of mica and of tinfoil. All the sheets, 1, 3, 5, 7, &c., of tinfoil are connected to one wire, and the alternate ones, 2, 4, 6, 8, &c., are connected to another wire, and correspond respectively to the two sheets of tinfoil shown in fig. 32.

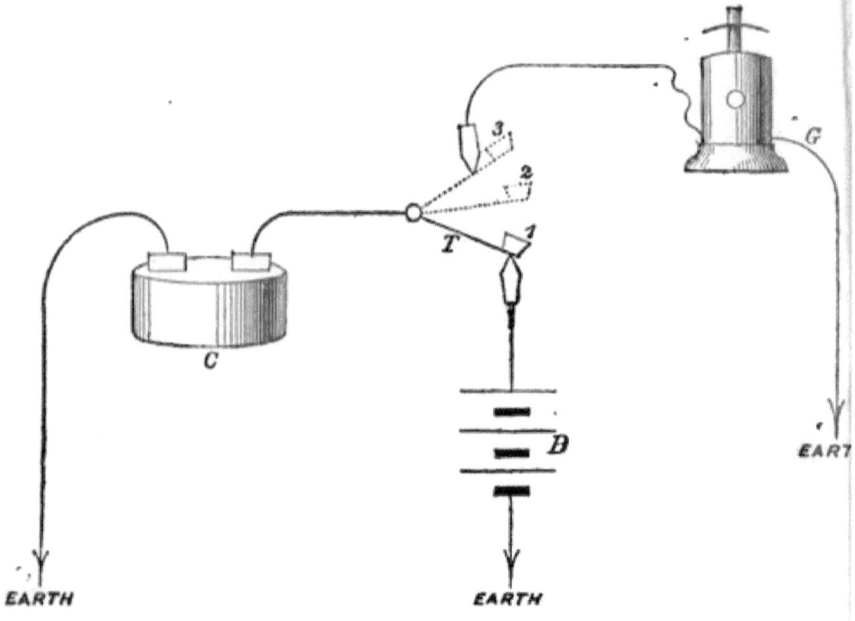

Fig. 33.

The sheets are all contained in the vulcanite box C, and the two sets are connected respectively to the two brass knobs on the lid. B is the battery, and the taps are replaced by the "key" T. The lever is connected to the condenser, and can be set in three positions.

In the position (1), when the lever presses against the lower stud, the battery charges the condenser.

In the position (2) the battery is disconnected.

In the position (3), when the lever presses against the upper stud, the condenser is discharged through the apparatus used for measuring the charge.

This consists of a Thomson's galvanometer as described on page 55. The first sudden swing of the needle (as indicated by the motion of the light spot) is proportional to the amount of charge, that is, to the pressure of the charging battery.

In order to determine what pressure a certain swing corresponds to, the experiment is first made with a Latimer Clark's standard cell (see p. 39) whose electric pressure is known, and then with the battery whose pressure is to be measured.

The ratio of the two **swings** is equal to the ratio of the two electric pressures.

Instance:—

A Latimer Clark's cell has a pressure of 1·42 volt. In a particular condenser experiment it gives a swing of 350 divisions of the galvanometer scale.

What is the electric pressure of a battery which, under the **same conditions**, gives **a** swing of 500 divisions?

Answer: It is $\frac{500}{350}$ of the pressure of the standard cell, i.e. it is $\frac{500}{350} \times 1\cdot42$ volts $= 2\cdot029$ **volts.**

In cases where the pressure to be measured is much greater, **say 50 or** 100 times greater than **that of the standard cell;** a branch wire called **a** "shunt" is connected across the galvanometer **poles, so that** only a small definite proportion, say $\frac{1}{100}$, of the electricity discharged passes through the galvanometer, and the other $\frac{89}{100}$ passes through the wire. The **deflection** is then only $\frac{1}{100}$ of what it would have been if **the whole** current had **passed** through the galvanometer, **and much** higher **pressures can** then be brought within **the range of** the instrument.

The **results calculated** when **a shunt is** used **must be** multiplied by the number representing the ratio of **the whole discharge to the fraction** of it which passes through the galvanometer.

61. The second method may be called the **direct or Pressure-gauge method.**

Water pressure may **be measured by means** of a pressure-gauge. The simplest **form of** water-gauge would be **a** cylinder and piston, **the piston** being pressed up by the water. Weights could then be laid upon the piston till the upward pressure of the water is just balanced. If the piston were exactly one square inch in area we should know that the weight which will just keep it down is equal to the pressure per square inch of the water.*

Electric pressure can be measured in the same way. If two conductors be placed near each other and connected respectively to the two poles of a battery, but not connected to each other, the current will wish to flow from one to the other, but will not be able to do so. They will then *attract* one another, and the attraction or the mechanical pressure forcing them together will be proportional to the electric pressure of the battery charging them. The mechanical pressure is practically very small, but can be measured by means of a fine spring. Thomson's and other absolute electrometers work on this principle, but the simplest instrument to illustrate it is Metchim's sine electrometer.

This instrument consists of two metal plates placed very near each other, one (*a*) (fig. 34) rigidly attached to the frame of the instrument, the other (*b*) hinged. The two plates can be set vertical or in any inclined position by a screw. When the plate *a* is vertical and the plates are not charged, *b* also hangs vertical. On the plates being charged *b* is attracted towards *a*, and forms an angle with

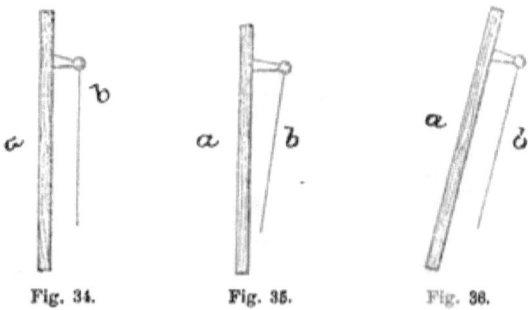

Fig. 34.   Fig. 35.   Fig. 36.

* Locomotive steam-engines work at a pressure of about 140 lbs. per square inch.

it, as in fig. 35. *a* is then tipped over by the screw, so that the weight of *b* pulling against the electric pull causes it to fall away from *a* till it is again parallel to it, as in fig. 36. Knowing the angle through which we have tipped the plate *a*, we know how much leverage we have given to the weight of *b*, and we know that the electric pull was equal to the pull due to that weight. Thus we can calculate the electric pressure.

62. The third method of measuring pressure may be called the **Small current** method, and is the one most useful in practice.

If we wish to measure the pressure of water in a reservoir we may drill a fine hole in it so that a little water may squirt out, the hole being so small that the quantity of water flowing out will not appreciably reduce the pressure.

The strength of the jet or stream which flows out will be a measure of the pressure.

The difference of electric pressure at any two points of a circuit, say the two sides of a lamp or the two poles of a battery, may be measured by attaching to those points the wires of any kind of galvanometer of high resistance. It must be of high resistance in order that the quantity of electricity flowing through it may not alter the pressure.

Any form of high resistance galvanometer where the needle is deflected by the current may be used, or we may use the method invented by Captain Cardew, and known as "Cardew's Voltmeter." This is, on the whole, the best voltmeter which I know, and it has the advantage of not being affected by moving masses of iron or magnets near it, and it gives the same indication for "direct" and for "alternating" currents.

It consists of a fine platinum-silver wire about six feet long, one end of which is fixed and the other takes a turn round the spindle of the indicating hand, and is kept tight by a spring. On the two ends of the wire being connected to two points of different electric pressure a current passes through it, whose strength depends on the pressure. This heats and expands the wire, and as it expands it turns the spindle and the hand.

The tube is made about three feet long, and the wire passes round a pully at the top of it. This is to save space.

The tube is made of two metals, so that its expansion rate is equal to that of the wire. Thus a change in the temperature of the room affecting the whole instrument equally does not affect the indicator.

Questions on Chapter IX.

1. What are the three principal methods of measuring electric pressure?

2. What is an electric condenser?

3. Give a mechanical illustration of the condenser method of measuring electric pressure, showing how water pressure might be measured by an analogous method.

Illustrate your description by sketches.

4. Describe a practical form of condenser.

5. Show how to measure electric pressures with it.

6. Show how water pressure can be measured directly.

7. Describe a simple form of water pressure-gauge.

8. Explain by what analogous method electric pressure can be measured.

9. Explain with sketches the principle of Metchim's electrometer.

10. Show how electric pressure can be measured by the small current method.

11. Give an illustration showing how water pressure could be measured by an analogous method.

12. Describe Cardew's voltmeter.

CHAPTER X.

EXPERIMENTAL MEASUREMENT OF RESISTANCES.

RESISTANCES are measured by comparing them with the resistances of certain standard coils, each of which has a resistance equal to so many "ohms;" that is, each of which is equal to the resistance of so many columns of mercury one square millimetre in section and 104 centimetres long.

63. Wheatstone's Bridge. We will suppose that we are furnished with a set of resistance coils, and that a wire is given to us of which we are to determine the resistance. We use an arrangement invented by Mr. Christie, and called "Wheatstone's Bridge."

We will first describe what may be called the "lecture model" of the apparatus, as it is a machine easy to understand, but inconvenient to work with, and then go on to describe the forms in practical use.

The lecture model (fig. 37) consists of a board on which is fixed a "diamond" of metal strips. At the four corners, A B C D, are binding screws, while in each side is a break with binding screws at each end of it.

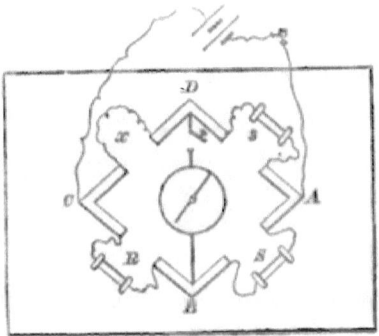

Fig. 37.

To two corners opposite to each other are connected the

battery wires; to the two other corners, those of the galvanometer. In the four breaks are put three known resistances, which we call S, *s*, and R, and the unknown one which we call *x*. We then vary the resistance R, and we shall find that at one particular value, the current of the battery produces no deflection of the galvanometer. When this is the case, we have, as we will prove immediately,

Ratio of *x* to R equals ratio of *s* to S;

from which *x* can be found by simple proportion.

64. Theory of Wheatstone's Bridge. To understand this, we must note the following direct deduction from Ohm's law :—

If a wire of uniform resistance be connected to a battery, the pressure varies regularly from one end to the other of the length—that is, at the middle point the difference of the pressure from that at either end is half that of the ends. At $\frac{1}{5}$ from one end the pressure differs from the pressure at that end by $\frac{1}{5}$ of the whole difference; and so on.

More generally, in any wire the pressure varies regularly along the resistance—that is, if there be a wire of 10 units resistance, and the pressure at one end is zero, and at the other is 100, the pressure at one unit from the first end will be 10, at two will be 20, at three 30; and so on

When the battery is in action, the current, on arriving at A (fig. 37), divides, as a stream might divide into two channels round an island, and part goes by the road ADC; part by ABC.

Let us now draw straight lines, ADC, ABC (figs. 38, 39), representing the resistances in the two courses, and let us draw vertical lines AL at the ends A, representing the difference of pressure between A and C.

Let AD represent the resistance *s*, DC the resistance *x*, then the total resistance of the branch of the circuit, containing *s* and *x*, is represented by the line ADC. Similarly the resistance of the branch containing S and R is represented by ABC.

The length of the line AL, which represents the excess of the pressure at A over that at C, is of course the same in both diagrams. **Draw lines LC** in each. Now, by what we have just stated—viz. **that the** pressure diminishes regularly—the pressure* at any other point in the circuit **can** be represented by the length of a vertical line drawn from the horizontal line at that point to the sloping line. The pressure at D, where one of the galvanometer wires is attached, is represented by the length DM (fig. 38).† In **a** similar way that at B, where the other wire is attached, is represented by BN (fig. 39).

Now, the effect of altering **the** resistance **R** will be **to**

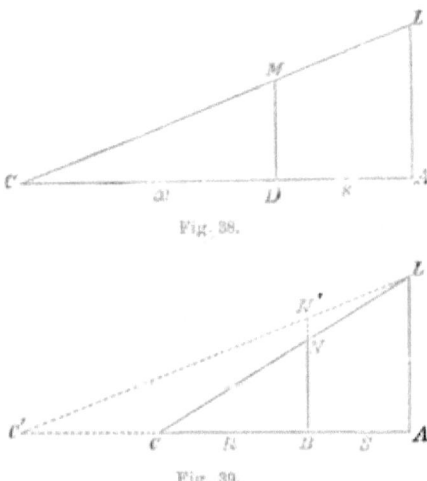

Fig. 38.

Fig. 39.

alter the pressure **at B; for,** suppose R increased so as **to bring** C to the **position C′,** then the fall of the pressure would be **represented by** the dotted line LC′, and the pressure **at B would** be represented by the length BN′. Let **us then vary R** until BN equals DM, *then, the ends of the galvanometer wire are at* the same **pressure,** *and there is no current through it.*

But as the height **AL is the same in both** triangles, the

* By "pressure" we mean "excess of pressure over that at C."
† The galvanometer circuit being broken.

heights at any other points D and B in the bases respectively can only be equal when the ratio of AD to DC—that is, of *s* to *x*, is equal to the ratio of AB to BC—that is, of S to R *—or, the galvanometer is at zero when

Ratio of *s* to *x* equals ratio of S to R.

But when two ratios are equal, the ratio of the first term of one to the first term of the other is equal to the ratio of the second term of the one to the second term of the other; therefore, when the galvanometer is at zero, we have

Ratio of *x* to R equals ratio of *s* to S;

or, *x* equals R, multiplied by the

Ratio of *s* to S,

which is written—

$$x = R\frac{s}{S};$$

Thus, when the galvanometer is at zero, *x* is known if the other three quantities are known.

The student will find it a useful exercise to prove for himself that it is unimportant at which pair of corners the battery wires are attached—that is, that interchanging the battery and galvanometer makes no difference.

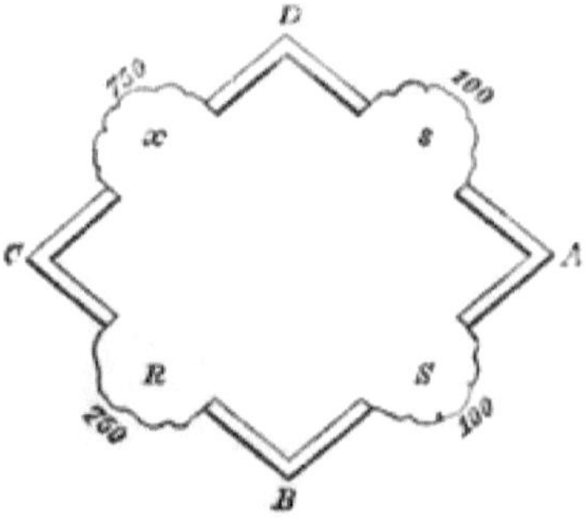

Fig. 40. 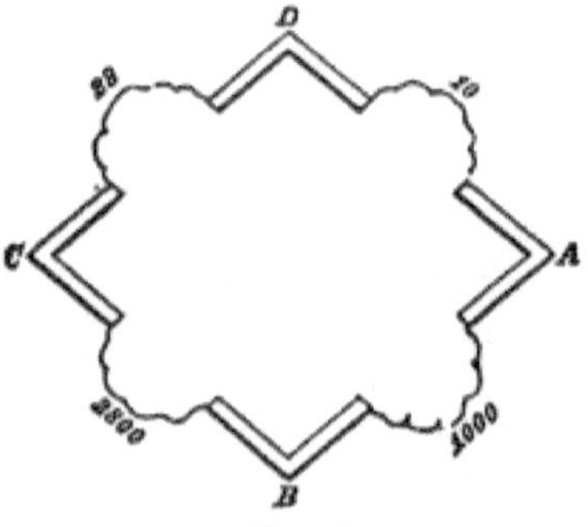   Fig. 41.

65. In practice the galvanometer and battery should be so

* The student is advised to test this with a scale drawing.

arranged that the two branches of the battery current encounter as nearly as possible the same resistance.

For instance, let us suppose S and s each equal 100, and R equal 750. Then $\frac{s}{S} = 1$; and, therefore, $x = R$, and our four branches will be as in fig. 40. The battery wires should now be attached to AC, and the resistance in each branch will equal 850. If they are attached to BD, the resistance in the one branch is 1500, and that in the other only 200. Now, however, suppose we have

$$S = 1000, s = 10, \text{ and } R = 2800.$$

We have, to find x,

$$x = R\frac{s}{S} = 2800\frac{10}{1000} = 28.$$

We must now attach the battery wires to BD, as in this case the resistances of the two branches will be 1010 and 2828; whereas, if we attach them to AC, the resistances will be 38 and 3800.

The objection to having a large difference in the branches is that nearly the whole current then passes through the circuit of least resistance, and heats it, thereby increasing the resistance on that side (for the resistance of a wire increases when it is heated), while, when the current is about equally divided, the increase of resistance due to the heating is about the same on both sides.

For convenience of calculation it is usual to make S a decimal multiple, or submultiple of s; then, when the balance is established by varying R, it is only necessary to multiply R by some power of 10, to find x.

66. Practical Forms of Wheatstone's Bridge. The form of bridge above described is never used except for lecture purposes. Two principal forms are used in practice. The one which is most used is called a "resistance box," and the other is called the "sliding bridge."

67. Sliding Bridge. This latter (figs. 42, 43) consists of a horizontal board, with a straight wire of high resistance stretched along one side, and a copper strip with gaps in it along the other. The diagram (fig. 43) should be compared with that of the lecture model (fig. 37)—the same letters indicate the same points in both.

Fig. 42.

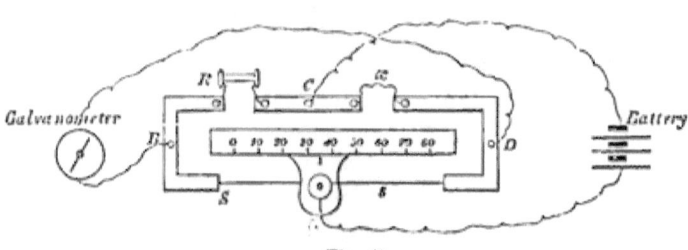

Fig. 43.

The connections are made as shown; R is not variable, but is chosen arbitrarily. A, one of the battery wires, is attached to a slider which slides along the resistance wire. Now the wire at one side of A is S, that at the other s. It is obvious that, *if the resistance of the wire is uniform*, the ratio $\frac{s}{S}$ will be equal to the ratio of the lengths of the wires on each side of the slider. A scale is fixed on the base, and if we suppose the equilibrium to be established with the slider at, say, 35, as in fig. 43, and suppose that R = 100, we shall have

$$x = R\frac{s}{S} = 100\,\frac{35}{45} = \frac{700}{9} = 77, \&c.$$

We see in this apparatus that it is not necessary to know the values of S and s, but only their ratio.

The apparatus is useful for some purposes, but it is not susceptible of any great accuracy, as its working depends on the assumption that a wire exposed to the air has a uniform resistance. As every particle of rust or scale, formed or rubbed off, and every scratch made by the slider affects the resistance, we see that the assumption cannot be held to be strictly true.

The slider usually carries a spring and vertical sliding rod, to which latter the battery wire is attached, so that contact is only made when wanted, by pressing down the spring.

68. Resistance Coils. A coil of wire of a known resistance is called a *Resistance Coil*. Resistance coils are usually made either of German silver wire, or of an alloy of silver with 33·4 per cent. of platinum, as the resistance of those materials varies very slightly with changes of temperature.

The two ends being fixed to massive copper rods, the wire, previously carefully insulated with two or more layers of silk, is wound double* upon a reel, as shown in fig. 44. It is usual to enclose the reel in a thin brass case, and imbed the wire in paraffin. By immersing the case in water, the wire can be brought to any desired temperature.

69. Bridge Resistance Box. A *resistance box* consists of a number of coils of different resistances arranged in the

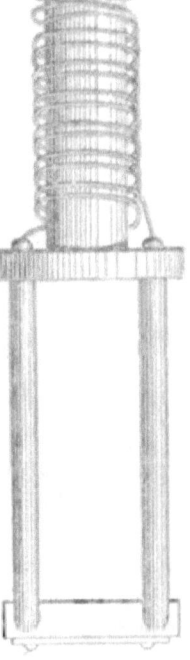

Fig. 44.

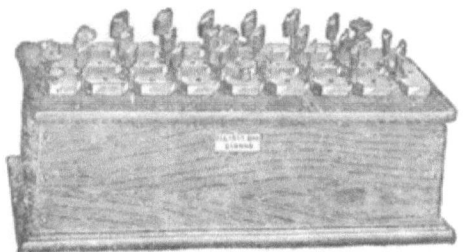

Fig. 45.

* The effect of the double winding is that there are always two equal currents in opposite directions close together. This entirely prevents any inductive effect on neighbouring magnets or wires.

following manner. The coils are all fixed to the under side of a slab of ebonite, which forms the lid of a mahogany box (fig. 45). On the top of the lid are a number of brass blocks, to each of which one end of each of two coils is connected, as shown in section in fig. 46. Thus, if a current is sent from A to B, it has to pass through all the coils. The ends of the brass blocks are, however, shaped as shown

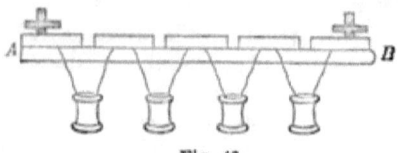

Fig. 46.

in the plan (fig. 47), and brass plugs fit in between them. When a plug is put in at any opening, the current only passes through the plug, and not by the coil; so that

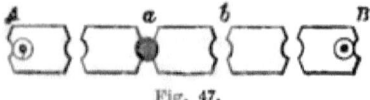

Fig. 47.

when a plug is put in, say at a, the total resistance is less by the resistance of the coil at a, or, generally, the resistance from A to B is equal to the resistance unplugged. The resistance of each coil is engraved on the lid near the plug-hole. Here, then, we have a means of varying the resistances without interfering with the connections.

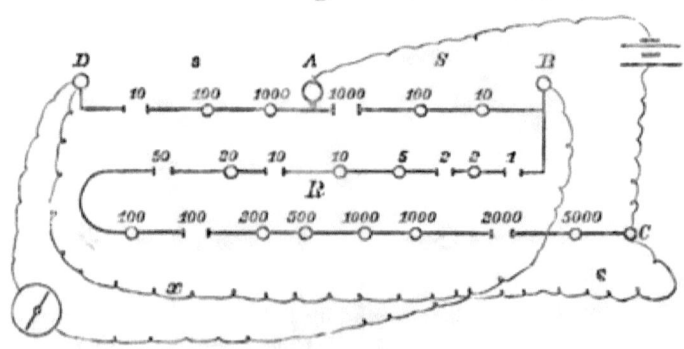

Fig. 48.

Bridge Resistance Box. 79

The coils in the bridge resistance box are arranged as in fig. 48, the numbers representing the number of units of resistance in each coil.

The reader is again requested to compare this figure with the picture of the lecture model, fig. 37, p. 71. The same letters are used for the same points.

We see that the coils are arranged in a continuous line, and binding screws inserted at certain intervals.

The first line contains two sets each of 10, 100, and 1000. These form the branches S and s. We see that they give the following values of $\frac{s}{S}$:—

$$\frac{10}{1000} = \frac{1}{100} \quad \text{also}$$

$$\left.\begin{array}{c}\frac{10}{100}\\ \frac{100}{1000}\end{array}\right\} = \frac{1}{10} \quad \frac{1010}{100}$$

$$\left.\begin{array}{c}\frac{10}{10}\\ \frac{100}{100}\\ \frac{1000}{1000}\end{array}\right\} = 1 \quad \frac{1100}{10}$$

$$\left.\begin{array}{c}\frac{100}{10}\\ \frac{1000}{100}\end{array}\right\} = 10 \quad \begin{array}{c}\frac{10}{1100}\\ \frac{100}{1010}\end{array}$$

$$\frac{1000}{10} = 100$$

Different forms of the same fraction are used for different resistances.

Thus, if we desired to have $\frac{s}{S} = 1$ we should make it $\frac{1000}{1000}$ if x were large, $\frac{10}{10}$ if it were small.

The longer part of the line of coils represents R on the bridge, and the arrangement of it is worth notice. There are only sixteen coils, and yet, by combining them, any number from 1 to 10,000 can be obtained. This is best seen by taking any number at random and trying to make it up. Generally speaking, we must use the large numbers first—that is, we must first unplug the largest number below the number we want.

x, the wire whose resistance is to be found, is attached to the binding screws C and D. If it should not be long enough to reach from one to the other, the length must be made up either by a wire of known resistance, or by a copper rod so thick that its resistance may be neglected.

70. Fractions of a Unit. We have stated that the ordinary resistance boxes give R a maximum value of 10,000 units.

When fractions of a unit are required, we get one or two places of decimals by making $\frac{s}{S}$, $\frac{1}{10}$, or $\frac{1}{100}$ respectively; but this can only be done when the resistances to be measured are not more than 1000 units for one place, or 100 for two places. A further approximation can be made by observing the deflections of the galvanometer. If it is a delicate one, it will seldom be found to go truly to the zero.

Suppose that the resistance is between 1221 and 1222, then we must have $\frac{s}{S} = 1$. When R equals 1221, the spot of light will move slightly in one direction, say 5 divisions to the left; and when R is 1222, it will move in the other direction, say 10 divisions to the right.

Then we shall approximate very closely to the truth if we say that the excess of the true resistance over 1221 bears the same ratio to its defect from 1222 as the deflection to the left bears to the deflection to the right; that is, that the true resistance is $1221\frac{1}{3}$.

71. Resistances of Different Substances.—*Specific Resistance.*—The specific resistance, R, of any substance is the resistance between two opposite faces of a cube of the substance when the edge of the cube is one centimetre.

When we know R for each substance, we can calculate the resistance of any wire by the formula—

$$\text{Resistance} = R \frac{\text{length}}{\text{cross section}}.$$

Specific Resistance.

The following **tables** of specific resistances of conductors and insulators are given in Professor Everett's "Units and Physical Constants," pp. 143—145 :—

As the resistances of all metals increase when they are heated, the "temperature correction" is given in each case.

Table of Specific Resistances, in Electro-magnetic Measure (at 0° C. unless otherwise stated).

	Specific resistance.	Percentage of variation for a degree at 20° C.
Silver, annealed	1521	·377
„ hard-drawn	1652	
Copper, annealed	1615	·388
„ hard-drawn	1652	
Gold, annealed	2081	·365
„ hard-drawn	2118	
Aluminium, annealed	2946	
Zinc, pressed	5690	·365
Platinum, annealed	9158	
Iron, annealed	9827	
Nickel, annealed	12600	
Tin, pressed	13360	·365
Lead, pressed	19850	·387
Antimony, pressed	35900	·389
Bismuth, pressed	132650	354
Mercury, liquid	96190	·072
Alloy, 2 parts platinum, 1 part silver, by weight, hard or annealed	2466	·031
German silver, hard or annealed	21170	·044
Alloy, 2 parts gold, 1 silver, by weight, hard or annealed	10990	·065
Glass at 200° C.	$2·27 \times 10^{16}$	
„ 250°	$1·39 \times 10^{15}$	
„ 300°	$1·48 \times 10^{14}$	
„ 400°	$7·35 \times 10^{12}$	
Gutta-percha at 24° C.	$3·53 \times 10^{23}$	
„ 0° C.	7×10^{24}	

The Ohm (or B.A. unit) = 10^9 or a thousand million absolute **units,** and thus to get the specific resistance in ohms of any substance we must divide this number in the table by a thousand million.

72. Galvanometer Shunts. For accurate determinations of resistances we must use the most delicate form of reflecting galvanometer. In cases where we do not approximately know the resistance there will, in the first few trials, be a considerable difference of pressure at the ends of the galvanometer wire; and, if the corresponding current were allowed to pass, the galvanometer would be damaged. To obviate this inconvenience, "shunts" are provided.

They are also useful for measuring, by the deflection of a galvanometer, currents so strong that, if sent directly through it, they would send the spot of light off the scale.

A shunt consists of a wire or coil of wire connected across the poles of the galvanometer. Part of the current goes through the coil, and only part through the galvanometer. By making the resistance of the coil have a certain ratio to that of the galvanometer a definite fraction of the current can be sent through the galvanometer instead of the whole of it.

Thus, if a shunt were such that $\frac{1}{10}$ of the whole current passed through the galvanometer, and the latter indicated one ampère, we should know that the whole current was 10 ampères. For the method of calculating these shunts see my "Electricity," 2nd edition, vol. i. p. 276.

QUESTIONS ON CHAPTER X.

1. Describe the "lecture model" of Wheatstone's bridge.

2. If a battery has a pressure of 10 volts, and the poles are connected by a wire of uniform resistance, and whose total resistance is 10 ohms, one ampère of current will go through it.

Show this from Ohm's law.

3. We will suppose that the wire is of such thickness that its resistance is one ohm per yard. It will then be 10 yards long.

If we connect one pole of the battery to earth where it is joined to the wire, the pressure at one end of the wire will

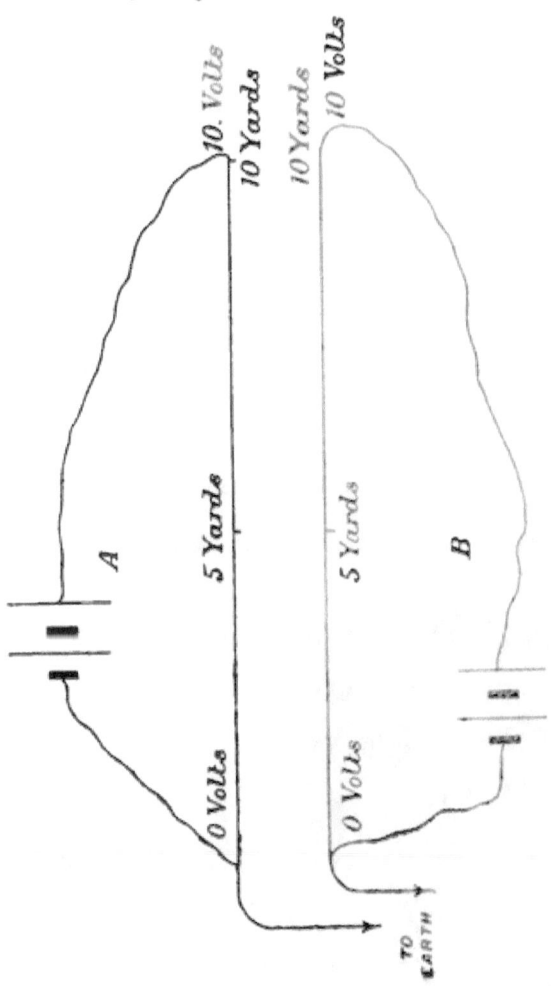

be zero, and therefore that at the other end will be 10 volts.

What is the pressure in the wire at (*a*) 1 yard, (*b*) 1½

yards, (c) 5 yards, (d) 7½ yards, (e) 8 yards 1 foot, (f) 9 yards 1 foot 6 inches, (g) 9 yards 1 foot 11 inches from the earth or zero end?

4. If we double the section of the wire so as to halve its resistance, keeping the length and the pressure of the battery constant,—

(a) How much current will go through?

(b) What will be the pressures at the various points mentioned in Question 3?

5. If we fit up two exactly similar batteries and wires, A and B, as described in Question 3, and connect (a) the the 5-yard point of A with the 5-yard point of B by a wire, will any current pass from A to B? (b) Will any pass if we connect 6-yard point of A with 4-yard point of B, and in which direction? (c) Will any pass if we connect the 8-yard point of A with the 8-yard point of B, and if so, in which direction?

5. We attach a wire from the middle point of A in the figure in Question 4, to the middle point of B, and we find no current passes in the cross wire. We then double the resistance of the whole wire B, still keeping it uniform. Will a current flow in the cross wire, and if so, in which direction?

6. Copy fig. 38, drawing it carefully to scale, then keeping the height A L constant, which represents the pressure at A, double the lengths x and b, which represent the resistances on the two sides of the point D respectively, join the new positions of C and L by the straight line C L. The height D M given by the point where the vertical line D M is cut by the line C M L will, as before, represent the pressure at D. (1) Will the height D M be increased or diminished or unaltered? What would have been the effect on the height DM if (2) we had doubled x without altering b, (3) if we had doubled b without altering x?

7. With two such figures, the height A L remaining constant, what relation between the x's and b's respectively is necessary to keep the height D M constant.

[In drawing these figures it is convenient to use the same letters in both, but to distinguish those of the second figure by a dash, thus x **may be the** x **of** the first figure, x' the x of the second figure.]

8. **Write** out § 75, "The Theory of Wheatstone's Bridge."
[**This** should be copied once or twice, and then written from memory.]

9. Explain with a sketch the sliding bridge.

10. What is a resistance **coil**?

11. How are coils fixed **into a** resistance box, **so** that they **can be cut in** or out of circuit as required?

12. **Let the** teacher copy out fig. 47 **on** the black-board, **and then** put chalk marks to certain coils, which are supposed to be unplugged when galvanometer is at zero. The class should then in each case calculate the value of the resistance x. If it is preferred to copy it once for all on drawing-paper, **the** coils unplugged can be marked by drawing-pins **stuck in.**

CHAPTER XI.

TELEGRAPHY.

73. The subject of telegraphy is so large that only the shortest outline of it is given here. Those who wish to obtain a practical knowledge of it are referred to special treatises on the subject.

To establish telegraphic communication between two places a *conductor,* which is always a metal wire, must be carried from one place to the other, and it must be *insulated* in order that the electricity may flow along the wire and not leak out at the sides. Telegraphic lines and instruments are different according as they are intended for overland or for submarine work.

74. Land Telegraphy. Land telegraph wires are of two kinds, "overhead" and "underground."

75. Overhead wires are most commonly used in England. It is unnecessary for me to give a detailed description of the method of fixing them as they can be seen running alongside all railways and elsewhere.*

The air being an insulator, it is only necessary to insulate the wires where these are attached to the posts. This is done by means of china insulators. The method of attaching them to the posts, and the wires to the insulators, may be seen by looking at them.

* Boys should be sent to look at the wires alongside the nearest railway, and at the next lesson the questions at the end of this chapter should be set to them.

Telegraphy.

Telegraph wires **are made** either of iron or **copper**. Overhead wires are generally iron. The thickness **of** wire required **is** fixed by the condition that **the** resistance of the line must not be unreasonably high. The resistance of iron is about seven times that of **copper,** and therefore an iron **wire to have** the same resistance as **a** copper one must have **seven times** the **section,** i.e. seven times the weight **per** mile.

It happens, however, that the cost of copper per pound is about seven times that of iron, and therefore the cost of an iron **or** copper wire of equal resistance **per** mile is about the **same.** Iron wire being much stronger **than copper** is preferred **for** overhead wires. In order to prevent **it rusting** away **it is** " galvanized "—that is, covered with **a thin coat**ing of **zinc.** The zinc **soon** rusts, but the rust or oxide of zinc is insoluble in **water,** and therefore protects the wire from further **action.**

76. Underground wires are covered throughout with gutta-percha, india-rubber, or **other** insulating material, and are laid **in** pipes **or** troughs under the streets. Copper is invariably used **for** the conductor, **as the** smaller wire requires less gutta-percha **to cover it.**

77. Single-needle telegraph receiver. This consists of a simple galvanometer **with the** needle vertical and stops to prevent it moving **too far either way.** A current passing through **the** coils in one direction moves **it** to the right one, in the other to the left.

78. Single-needle sending key. This consists **of a lever** which, **on** being moved **to** the right or left, connects **the positive or** negative **end of the** battery respectively to **the wire.**

A **sending and** receiving instrument **are** placed **at** each end of the line, so that messages can be **sent** either way, and they are **so** connected that the **current always** goes through both galvanometers, and the **sender can therefore see** on his instrument **the message** which **he is sending** to the other end.

79. Earth plates. If we wished **to send** signals through a water-pipe by starting and stopping the flow of water, it would be necessary for the pipe to be open at the ends, and to dip into reservoirs of water so large that no flow through the pipe could alter their level. Similarly, the ends of the electric wire must be open and be connected to the earth, which may be regarded as a reservoir in which differences of electric pressure instantly disappear.

To make connection with the earth, either large plates of metal are buried in a damp place and the wires connected to them, or the wires are connected to the nearest system of water-pipes.

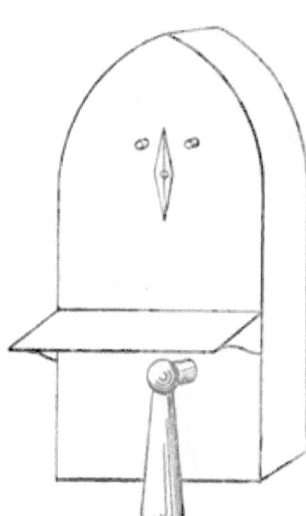

Fig. 49.

The diagram (fig. 50) shows the connection of a simple telegraph circuit.

On the battery poles $a\ b$ being connected to the line wires $a'\ b'$ respectively, a current will flow through the line in the direction of the arrows.

On their being connected across, viz. a to b' and b to a', a current will flow in the other direction.

80. Morse printing instrument. This is an instrument for *recording* the messages sent. The receiving instrument consists of an electro-magnet, over which is a lever carrying a pen. The lever is held up by a spring, and a long ribbon of paper, moved by clockwork, passes under it.

On a current being sent through the magnet the pen presses on the paper and makes a mark. If the current only lasts for an instant a dot (·) is made, but if the contact key is held down so that the current lasts a little longer a dash (—) is made as the paper moves under the pen. Combinations of these dots and dashes form an alphabet.

The following is the code in general use :—

A • —
B — • • •
C — • — •
D — • •
E •
F • • — •
G — — •
H • • • •
I • •
J • — — —
K — • —
L • — • •
M — —
N — •
O — —
P • — — •
Q — — • —
R • — •
S • • •
T —
U • • —
V • • • —
W • — —
X — • • —
Y — • — —
Z — — • •

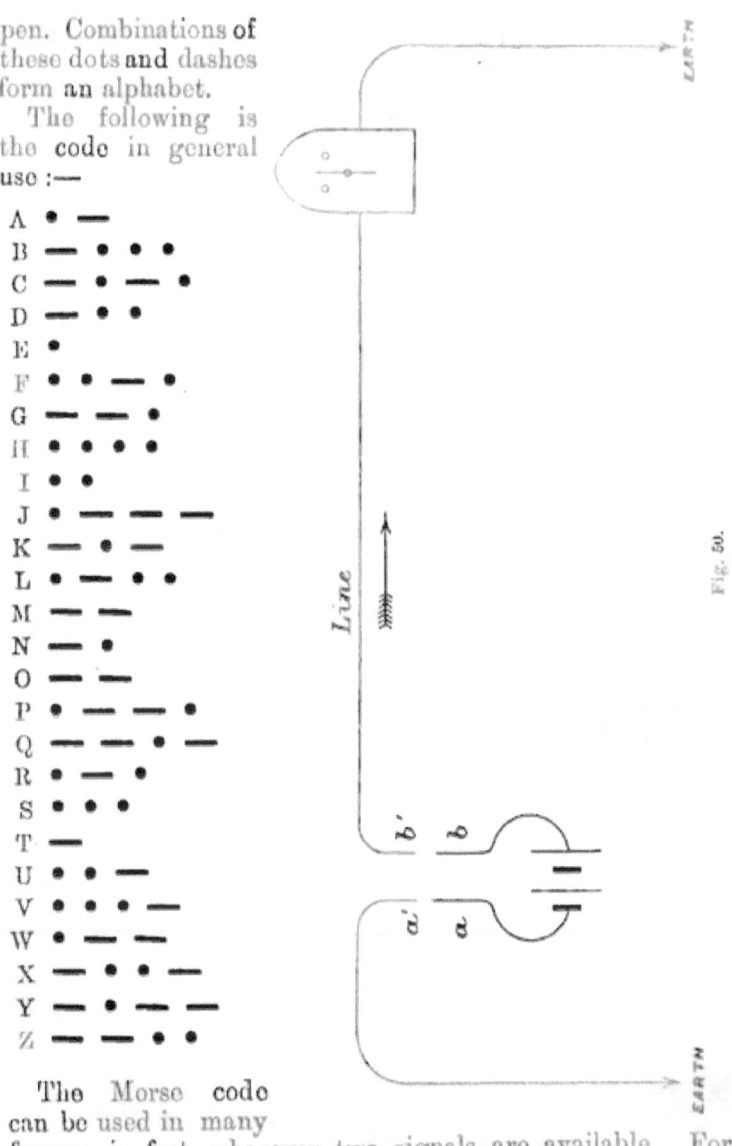

Fig. 50.

The Morse code can be used in many forms; in fact, wherever two signals are available. For instance, steamships can talk with their whistles, a long whistle representing a dash, and a short one a dot.

Heliographic signalling, so much used in war-time, consists of long and short flashes of sunlight produced by slight motions of a mirror.

Two persons skilled in the code can talk secretly in company, by agreeing, for instance, that a motion or a tap on the table by the first finger shall mean a dot, and by the second finger a dash. Or winks of the right and left eye can represent dots and dashes respectively.

Many other delicate instruments are in use, some of which print the message in ordinary type.

81. Submarine telegraphy. For telegraphing between places separated by sea the conductor has to be much more carefully protected than is necessary in the case of a land-line.

The following is the general construction of a submarine cable: The conductor consists of a strand of fine copper wires, which is much less likely to be broken than a solid wire. The strand is about one-eighth of an inch in diameter. This is thickly coated with gutta-percha, which completes the electrical portion of the cable, and the gutta-percha-covered wire is called the "core." The rest of the coverings are solely to protect the gutta-percha from mechanical injury. The core is next covered with white tape, folded round it longitudinally. Next a brass tape is wound spirally round it. This to protect the core from the attacks of boring insects called "teredos." In seas where teredos do not exist this brass tape is omitted. Next a black waterproof tape is wound spirally round it. Then a number of strands of yarn are wound on, giving the cable the appearance of a stout rope. The cable is then over-spun with a number of strong steel wires which completely encase it. Then two more black tapes are wound on, one in a right-handed spiral and one left-handed. These tapes are tarred as they are put on. The whole of the processes above described are carried out by automatic machinery. The cable is then complete, and is coiled in great tanks filled with water, where the insulation is tested.

One end of a battery being connected with the **water in the tank, the other end is connected** to the conductor of the cable, **and a** difference **of electric** pressure is **produced** tending to send **a** current **from the** conductor to **the water.** If the insulation is perfect no current passes, **but** if there is the most minute pin-hole in the gutta-percha a current will pass which can be detected by the reflecting galvanometer, and the "fault" has to be found and repaired. This testing is carried on continuously through the whole process of manu**facture, and** through all the subsequent **processes** of laying.

A deep-sea cable **when** completed is about **one** inch in **diameter, but** the shore ends, i.e. **the portions** which lie **in** shallow water, and are liable **to be fouled by** anchors, &c., are made much thicker by **means of an extra** thickness of iron **wire.**

The **cable** remains in the tanks until the ship which **is to** lay it is ready, **when** the latter is brought alongside **of** the wharf, and the **process of** loading commences. A cable factory must **always be so** situated on the river-bank **that a** ship of 4000 or **5000 tons can come up** close to **the factory.** In the ship **are two or three great** circular **tanks, perhaps 35** feet in **diameter and 25 feet deep, and** a system **of pulleys and** guides **is run from the tanks in** the factory **to the tanks on** board **the ship. The cable is run** out by a small steam winch **at the rate of some three miles** per hour, and is coiled in the tanks by **men who are, of course,** provided with soft boots that they may **not** injure the portions already **coiled.** When **the** cable is loaded **the tanks are** filled up **with water, and the ship is** ready to **proceed to the starting-place.**

One end **of the cable is passed** through an arrangement **of brake-wheels,** which regulate the pace at which **it will be paid out, and is** passed **over the stern of** the ship and brought **on shore.** The **ship then starts, and** steams slowly onward, **paying out cable as it goes until the line is** completed.

Deep-sea telegraphy **commenced in 1864, when** the Telegraph Construction and Maintenance Company laid the first successful **cable to** America.

Since then about 100,000 miles of submarine cable have been laid in different parts of the world, their total value being about 20,000,000*l.* sterling.

82. Repairs of submarine cables. Cables are liable to certain accidents (one was broken by the Java earthquake, and one by a whale who got entangled in it and was killed in his struggles), and also suffer from old age after a certain number of years. When a fault occurs it has first to be found, i.e. its distance from the shore has to be measured.

This is done by measuring resistance by the Wheatstone's bridge process, which we have already described.

As the simplest kind of fault, let us suppose that the cable is cleanly broken, and that we have breach of continuity and "dead earth." We know the total length of the cable and its resistance before it broke down.

Suppose the sound cable is 3000 miles long, and had a resistance of four ohms per mile; that would be a total resistance of 12,000 ohms, i.e. the current leaving one, say the English, end of the cable has to pass through 12,000 ohms resistance before it can get to earth at the other, say the American, end.

Now, suppose the cable broken. The current which enters at the English end goes to earth at the fault, i.e. only passes through the resistance of that part of the cable which is between the English end and the fault. Let us suppose that the bridge test shows us that the resistance is now only 4000 ohms, then, as the resistance is four ohms per mile, we know that the fault is 1000 miles from the English end.

Few faults are, however, as simple as the one we have described. They generally consist of slight leaks, which, while injuring the communication, have a considerable resistance of their own, and which therefore require more complicated methods of testing in order to find how much of the resistance observed is due to the fault itself, and how much belongs to the wire leading up to the fault.

One method is to test from both ends. Now with a cable such as we have described, of 12,000 ohms resistance,

3000 miles long, and with a complete break in it 1000 miles from England, a test taken from England would give 4000 ohms as the resistance from England to the fault, and a test taken from America would give a resistance of 8000 ohms as the resistance from America to the fault, and the two together, 8000 + 4000, make up the 12,000 ohms resistance of the cable.

Now, suppose that we have a much smaller fault in the cable, a fault which has a high and unknown resistance.

Say that the test from England gives 5000 ohms as the resistance of the wire from England to the fault, plus resistance of fault, and we do not know how much of this is fault and how much wire.

Suppose that the test from America gives 9000 ohms as the resistance of the wire between America and the fault, plus that of the fault.

Let us call R_E and R_A the resistances of the English and American portions of the wire respectively, and R_F the resistance of the fault. The total resistance of the cable is of course $R_E + R_A$, and we know this is 12,000 ohms.

We have then—

$$R_A + R_F = 5000$$
$$R_E + R_F = 9000.$$

Adding them together we have—

$$R_E + R_A + 2R_F = 14,000.$$

Subtracting the known value of—

$$R_E + R_A = 12,000$$

we have—

$$2R_F = 2000 \text{ ohms,}$$

or—

R_F, the resistance of the fault = 1000 ohms.

But—

$$R_E + R_F = 5000 \text{ ohms,}$$

and if we substract the value of R_F we have—

$$R_E = 4000 \text{ ohms;}$$

or the fault is 4000 ohms, or 1000 miles, from the English end.

I have here only outlined two of the simplest methods of testing for faults. Innumerable refinements are in use in practice.

The reader who wishes to study this subject further is referred to Kempe's "Handbook of Electrical Testing."

83. Repairs. When the fault has been localized, a ship is sent to the place, provided with grappling ropes and hooks and a small quantity of cable.

Taking up a position a little to one side of the cable and opposite a part of it a mile or two from the fault, she drops her grapnel and, moving across the cable, draws it along the bottom till the cable is caught. It is then drawn up to the surface, and the broken end is attached to a buoy and there left.

The ship then goes to a point a mile or two on the other side of the fault, and picks up the other end.

To this she attaches the end of the spare cable which she has on board and steams back, paying it out till she arrives at the buoy which holds up the first end. She then brings both ends on board, i.e. the first end and the end of the new piece, and having connected them throws them overboard, and the job is complete.

Immediately before connecting, the chief electrician on the ship telegraphs to both shore-stations, "Am about to make final splice," and an hour or so later the electricians at the two ends (i.e. say in England and America) will find that they are in communication with each other, but no longer in communication with the ship.

Repairs in two miles depth of water are now made with almost perfect certainty.

I cannot in this book describe in detail the signalling apparatus used in submarine work. The chief difference between them and those used for land-work is, that the submarine instruments are much more delicate, and allow much smaller currents to be used.

The reflecting galvanometer is always used as the receiving instrument.

Questions on Chapter XI.

1. Give a sketch and short description of an overhead telegraph line.
2. How are the wires insulated from the posts?
3. About how many posts are required per mile run?
4. Describe generally the single needle telegraph receiver.
5. Sketch the connections of a simple telegraph line.
6. Describe the Morse printing instrument.
7. Describe the general construction of a submarine cable.
8. Give an outline of a simple method of testing for faults in a submarine line.
9. Describe the general process of repairing a cable.

CHAPTER XII.

THE TELEPHONE.

In the telegraphic instruments which we have described, simple signals are transmitted through the wires, and it is agreed that certain signals and combinations of signals shall represent certain letters, and so messages are spelt out.

84. By means of **The Telephone** articulate speech is transmitted.

The telephone is an American invention, and has been claimed by Professor Graham Bell, Mr. Edison, and others.

Fig. 51 shows the construction of Bell's first telephone.

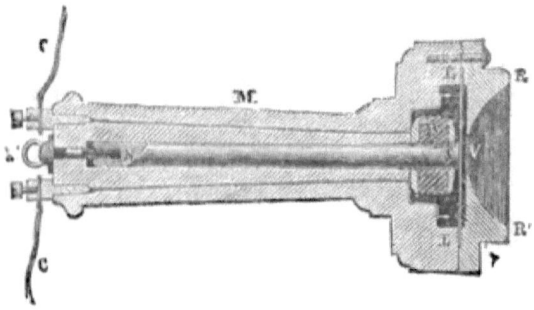

Fig. 51.

It consists of a steel magnet S N, round one end of which is wound a coil of fine wire B. The magnet and coil are enclosed in a wooden tube M, one end of which, R V R, is of considerably greater diameter than the magnet.

Across the **wide end** of the tube **a** diaphragm of thin sheet iron L L **is fixed,** which just does not touch the pole of the magnet.

When the instrument is spoken to, the iron plate vibrates in time with the sound vibrations.

As it moves it causes temporary alterations in the magnetism of the steel magnet, and these in turn induce periodic currents in the coil of wire.

The induced currents are **conveyed** along **the** telegraph line C C, and received **in a** similar telephone **at its** other end.

They travel **round** the coil of wire in **it, and** cause temporary changes in **the** magnetism of the **steel** magnet.

Owing to these changes, the force with which the iron plate is attracted **varies,** and the latter is caused to vibrate in time with the vibrations of the plate of the sending instrument.

The plate, as it vibrates, sets **the** air in motion and reproduces exactly not only the note but *the words spoken* **into** *the sending instrument.* The voices of different speakers **can** be recognized **even at a** distance of many miles.

When a *steady* battery current is sent through a telephone, no sound is produced, but **every variation** of **the** current causes a loud noise.

85. The Hughes microphone. When at any **point** in a circuit carrying **a** battery-current there is **an** imperfect contact, any change in the goodness of the contact will produce a change in the current and cause a sound in **a** telephone included in the circuit.

Professor Hughes has discovered that, when the imperfect contact co**nsists of** two pieces **of** carbon lightly pressed together, **variations in** the current are caused **by** the very smallest sound occurring near the carbon.

The *microphone* consists **of** two or more pieces of carbon lightly pressed together. **A** telephone and a battery **are** included in circuit **with** it.

The lowest whisper spoken near the microphone is loudly reproduced in the telephone.

We **see** that **in the** simple form of the telephone the

voice has to produce the currents which pass along the wires. Their strength is then limited by the strength of the voice.

By the use of a microphone we are enabled to produce the current by means of a battery, and only use the voice to cause pulsations in it, and are thus able to obtain much more powerful effects than with the telephone only.

86. The Gower-Bell telephone. This is the latest

Fig. 52.

form of telephone, and, I think, the one which is in most

The Telephone.

general use. Fig. 52 shows its general external appearance. Fig. 53 is a diagram of the connections.

The case of the instrument is closed at the top by a sounding-board, consisting of a sheet of thin wood about eight inches by six inches.

87. The sending apparatus. To the under side of this sounding-board is fixed a microphone (fig. 53), consisting

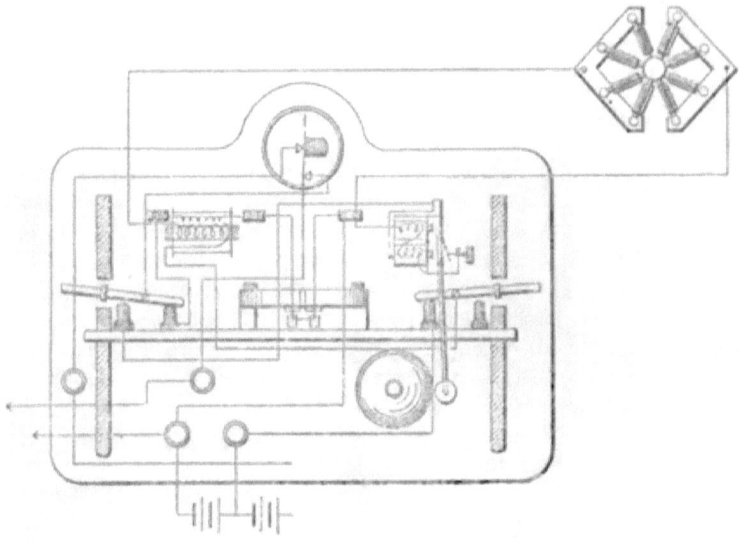

Fig. 53.

of a number of pieces of carbon lightly fixed together. A battery is connected so that when the sounding-board is reverberated by sound the variations of contact between different pieces of carbon cause pulsations in the current.

This completes the sending part of the apparatus. It will be understood that each instrument contains a sending and receiving apparatus, so that it can both speak and hear. For convenience the connections are so made that each sending apparatus causes both speakers to speak, i.e. both the one at the distant station and the one close by it.

88. The receiving apparatus. The current is received

in a small *induction coil*,* which consists of a little electro-magnet with another coil of wire wound over it. The varying currents in the electro-magnet, called the *primary coil*, cause pulsations in its magnetism. These again induce currents in the outer, or, as it is called, the *secondary coil*. These secondary currents pass through another electro-magnet, whose pole is fixed close to a thin plate of sheet-iron similar to that in the simple Bell telephone (fig. 51), and setting it in vibration cause it to speak.

In order that the speech may be more clearly heard, the iron diaphragm is let into one side of a closed chamber from which lead two flexible tubes which can be placed to the ears.

89. The call bell. An ordinary electric-bell † is provided to call attention. In order that the same wire may be used for both bell and telephone, a horizontal lever is arranged so that when it is pressed down the bell is connected and the telephone disconnected, and so that the reverse takes place when it is lifted. A spring presses it up, but the flexible hearing-tubes are hung on it when the telephone is not in use, and so keep it down. On the bell ringing the person addressed lifts the tubes and puts them to his ears, and by the act of so doing disconnects the bell and connects the telephone.

Telephones are chiefly used for distances under thirty or forty miles, but a telephone-line is in regular use between London and Brighton, and speech has been transmitted over 400 miles as an experiment.

90. Telephone exchanges. In London and other great towns, the telephone companies arrange that every subscriber shall be able to speak at will to every other subscriber.

For this purpose an office, called an "exchange," is opened in a central situation, and all wires from the subscribers' houses or offices are brought to it. An alphabetical list of all the subscribers' names and addresses is printed and sent to each of them, and to each name a distinguishing number is attached.

* See page 162. † See page 102.

We will suppose that Smith, No. 4050, wishes to speak to Brown. He looks in his list and finds that Brown is No. 2756. He then calls the exchange, **and** says, "4050 wishes to speak to 2756." The exchange then connects the wire 4050 to the **wire** 2756, which puts Smith and Brown into direct communication.

Special instruments and distributing boards are used in **the** exchange to make the communication.

QUESTIONS ON CHAPTER XII.

1. **Describe,** with sketches, the simple **Bell** telephone.
2. **Describe the** microphone.
3. **Describe, with** sketches, **the** Gower-Bell telephone.
4. **What is a telephone** exchange?

CHAPTER XIII.

ELECTRIC BELLS.

91. Electric bells are now very widely used, both as adjuncts for telegraph instruments and for domestic purposes.

Fig. 54 shows the construction of an ordinary electric

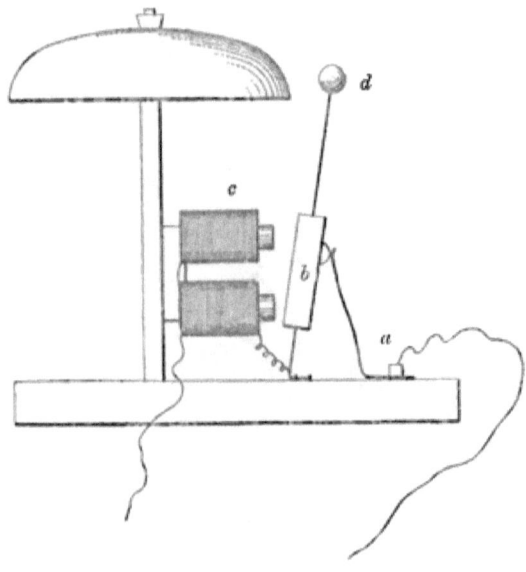

Fig. 54.

bell. The current passes in through the spring (a) down by the back of the armature (b), and so to the electro-

Electric Bells.

magnet (*c*). The magnet then attracts the iron armature (*b*), and the hammer (*d*) strikes the bell. In moving forward, however, it leaves the spring (*a*) and breaks the contact, and the other spring (*e*) causes the hammer to fly back. On its so doing contact is again made at (*a*), and the hammer again strikes the bell, and so a continuous ringing is kept up.

The bell is rung by pressing a button, which brings two metal springs together and so closes the circuit.

In a house or hotel there is a contact-button in every room, so arranged that pressing any one button rings the bell.

In order that the person summoned by the bell may know from which room the call has come, an indicator-board is placed near the bell. Behind this board are a number of electro-magnets, each in connection with one contact-button. On the button being pushed the bell rings, and the particular magnet in connection with that button pulls over a lever and causes a red disc to appear behind a corresponding hole in the board.

Fig. 55 shows the general connections.

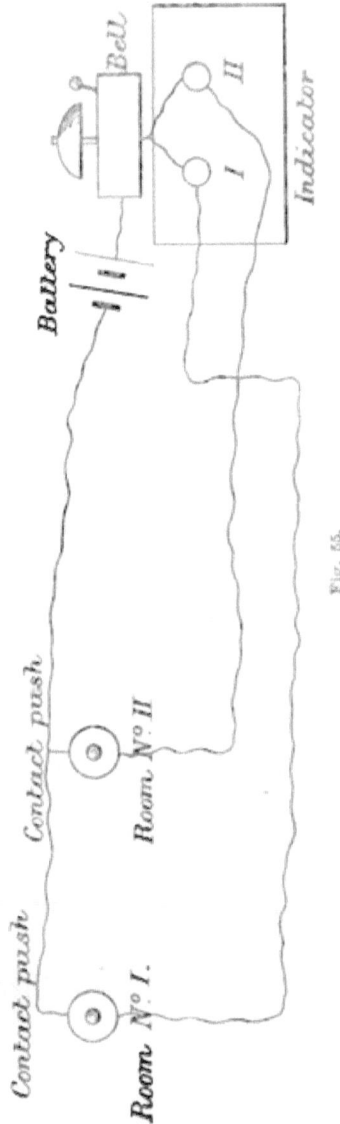

Fig. 55.

Questions on Chapter XIII.

1. Describe and sketch an electric bell.
2. Describe and sketch the connections of an electric bell system with indicators.

CHAPTER XIV.

ELECTRIC LIGHTING.

92. Introductory. In all systems of artificial lighting of whatever kind, the light is produced by the incandescence or glowing of solid particles of matter. The heat required to produce this incandescence is produced in various ways. Our ordinary coal-gas consists of a combustible gas, richly charged with very small solid particles of carbon, which, being made white-hot by the combustion of the gas, glow and produce the light required.

If these solid particles are removed, the combustion of the gas produces no light. If, for instance, we mix with the gas a sufficient quantity of air to oxidize and consume the carbon particles, we obtain a flame hotter than the ordinary gas flame, but giving no light at all. A burner contrived specially to mix the right proportion of air with coal-gas, and known as the "Bunsen burner," is much used for cooking, and for other purposes where heat without light is required.

Again, if we burn pure hydrogen gas we shall obtain a flame giving great heat, but no light, because the hydrogen does not contain any solid particles. If, however, we introduce a little spiral of fine platinum wire into the flame, the heat of combustion will make the wire white-hot, and it will glow and give light. If, instead of the thin platinum wire, we use a thick one, it will only become perhaps just red-hot, and although the same quantity of heat is being used, for the same quantity of hydrogen is being burned, yet the wire gives very much less light than before.

If now, instead of allowing the hydrogen to burn in the air where the oxygen, with which it is combined, being diluted with nitrogen, is diffused over a considerable space, we supply it with pure oxygen, the heat produced is concentrated in a much smaller space, and the temperature of the flame is consequently much higher. If the mixed gases are supplied under pressure, the size of the flame is still further reduced, and it can be concentrated on a very small portion of the surface of any solid body against which it may be directed. When the "oxy-hydrogen jet" is directed against a cylinder of lime, it raises a portion of its surface to a very high temperature indeed, and the heated lime gives off an intense light. This arrangement is well known as the "Lime-light."

Now, in all these arrangements we get a certain definite quantity of heat from a given quantity of fuel consumed, and this amount is the same in whatever way the combustion takes place. The same total quantity of heat is produced by the combustion of a cubic foot of hydrogen, whether it burns with a large flame in air, or whether it burns in an oxy-hydrogen blow-pipe.

The amount of light, however, which is produced by the expenditure of a given quantity of gas, or by the production of a given quantity of heat, depends entirely on the way that heat is applied.

Let us consider, for instance, the heat produced by the combustion of a cubic foot of hydrogen burnt in ten minutes. This heat may be expended in boiling a certain quantity of water, in which case it produces no light at all; or it may be employed in heating a thick platinum wire to dull redness, in which case it produces a little light, or it may be used to heat a thin wire to whiteness, producing a considerable light; or, finally, by means of the oxy-hydrogen jet, it may be employed in heating a piece of lime to intense whiteness, and produce the lime-light.

We see, therefore, that to produce a light we must heat a solid body to incandescence. To produce a given light with the smallest possible expenditure of heat (that is of fuel and cost) *we must concentrate our heat on a solid body of the* smallest possible **size, so** that that **may be raised** to the highest possible temperature.

Solid bodies can be rendered incandescent and made to give **light** by heat produced otherwise than by combustion. In the **old** " Flint-mill," used by miners before the invention of the safety-lamp, heat was produced **by** means of the energy applied by the man who turned **the** handle and caused **a** flint and steel to be continually knocked together. **The heat** caused the incandescence **of** the particles of flint **knocked** off, and the stream of sparks gave a certain quantity of light.

Solid bodies can also be made **hot** by the passage of **a** current **of** electricity through them; and it is by this heating that electric light is produced.

Questions on Chapter XIV.

1. **What is** the necessary condition for the transformation of heat into light?

2. **Illustrate how the same** quantity **of** heat may produce different quantities of light.

3. **What is the condition that a** given quantity of heat may produce the greatest possible quantity of light?

CHAPTER XV.

ON THE CONVERSION OF ELECTRIC CURRENTS INTO HEAT.

93. Heat produced by friction of water. (1.)* Let us suppose we have a long water-pipe of large bore, bent round so that its two ends dip into the same cistern at the same level, and let a force-pump be connected to one end. We see that by a very small power we can cause a stream of water to flow round it. The only force opposing the motion will be the friction of the water in the pipe. To overcome this friction, however, a certain quantity of heat has to be expended in the steam-engine working the force-pump, and, by the friction, the sides of the pipe and the water will be more or less heated.

(2.) If the pipe is of small bore then more work will have to be expended to send a given stream of water through the pipe, and the friction being greater the pipe will be more warmed.

(3.) We see, then, that the stream of water has given us a means of taking heat from the engine-fire, and conveying it to a distance, namely, to every portion of the sides of the pipe.

(4.) As long as the pipe is of the same bore throughout, the friction will be the same at all parts, and the heating will be uniform all along the pipe. If, however, we were to cut our pipe at one place, and interpose a spiral of very fine tube, a great deal of friction would be concentrated at one spot, and instead of the heating being uniform all along the pipe, by far the greater portion of the total heating would take

* Compare the numbered paragraphs with those having corresponding numbers on page 110.

place in the spiral. **Thus this** arrangement of the stream has given **us a** means **of** taking heat from the engine-fire, and conveying **it** to any one place we like at **a** distance, namely the place where we have put the spiral.

(5.) We see that we have expended *mechanical* **work** in forcing **a** *current* of water through a pipe offering *resistance* to the flow; which pipe, by its resistance, has reconverted a portion of the current into heat. The distribution of the heat depends **on** the distribution of the resistance. **When** the resistance is evenly distributed all along the pipe, **the** pipe is evenly warmed. When the greater portion of the resistance is concentrated **at one** spot, the greater portion of the heat **is** produced at that spot.*

(6.) We must particularly note that the heat used **has** been expended in forcing a current of water through **a** resistance, and **not** in producing the water itself; and that if **a** pipe could be made without friction, and thus offering no resistance to the flow, then that a stream of water, however strong, when **once** started would go on flowing round **the** pipe for ever without the expenditure of any work at all.†

94. Electric current. Now we know that in "conductors" a current **of** electricity can **be** made to flow in the same way as our current **of water** round the pipe. As we have already stated, **no substances** are quite perfect **either as conductors or insulators, the best** conductors offer some resistance **to the flow, and the best** insulators allow **a** little electricity to pass through them; but for **our present** purpose the metals and carbon among solids, **and intensely** heated or highly rarified **air** and gases may **be classified** as conductors, and all other solids, and air and gas at ordinary temperatures and pressures, may be called insulators. With the **conducting** and insulating powers of liquids **we have at present nothing** to do.

Conductors **differ greatly** among themselves in the facility

---

* For the purpose **of this illustration** we consider the heat to stay **in** that portion of the **pipe in which it is** produced, and **not** to be carried away by the water.

† Newton, Lex. I

with which they conduct electricity. A platinum wire, for instance, offers between five and six times the resistance to the flow of electricity as a copper wire of the same length and diameter. Also the resistance of a given length of a given wire is greater when the wire is thinner, being inversely proportional to its cross section. With the same cross section it is directly proportional to the length.

95. Heat produced by electric current. (1.) * Let us now suppose that by means of a steam-engine turning an electric generator we are forcing a current of electricity through a long copper wire of large diameter. The only force opposing the flow will be the resistance of the wire. To overcome this a certain quantity of heat has to be expended in the steam-engine working the electric generator, and, by the resistance, the wire will be more or less heated.

(2.) If the wire is of smaller diameter the resistance will be greater, more work will have to be expended to send a given current through it, and more heat will be produced in the wire.

The relative amounts of work expended and heat produced in sending a current of electricity through a thick and a thin wire of the same length and material are inversely proportional to the cross sections of the wires.

(3.) We see that the electric current has given us a means of taking heat from the engine-fire and conveying it to a distance, namely, to every portion of the wire.

(4.) As long as the wire is of the same diameter and of the same material throughout, the resistance will be the same in all parts, and the heating will be uniform all along the wire. If, however, we cut our copper wire at one place and interpose a spiral of very fine platinum wire, a great deal of resistance will be concentrated at one spot, and instead of the heating being uniform all along the wire, by far the greater portion of the total heating will take place in the spiral.

Thus this arrangement of the electric current has given us a means of taking heat from the engine-fire and conveying

---

* Compare the numbered paragraphs with those having corresponding numbers on page 108.

it to any one place we like at a distance, namely, the place where we have put the spiral.

96. Analogy between electric current and water current. (5.) In the electrical case also we see that we have expended *mechanical work* in forcing a *current* of electricity through a wire offering *resistance* to the flow; which wire, by its resistance, has reconverted a portion of the current into heat. The distribution of the heat depends on the distribution of the resistance. When the resistance is evenly distributed all along the wire, the wire is evenly warmed. When the greater portion of the resistance is concentrated at one spot, the greater portion of the heat is produced at that spot.

(6.) We must particularly note that the heat used has been expended in forcing a current of electricity through a resistance, and not in producing the electricity itself (whatever that may be); and that if a wire could be made offering no resistance, then a stream of electricity, however strong, when once started would go on flowing round the wire for ever without the expenditure of any work at all.

97. Ampère's theory of magnetism. An electro-magnet consists of a bar of soft iron surrounded by a coil of copper wire. When an electric current is sent round the wire the iron bar becomes a magnet. The copper wire offers resistance to the flow and becomes heated, and therefore work has to be expended at the generator to keep up the flow.

In a permanent steel magnet, according to the theory of Ampère, the magnetism is produced by the continuous flowing of electric currents in channels of no resistance which surround the molecules.

We may therefore regard a permanent steel magnet as an electro-magnet surrounded by a wire of no resistance, in which the current having once been started, by whatever process of magnetization has been adopted, continues to flow eternally without producing any heat or requiring any heat to maintain it.

98. Theory of electric lamps. When at the place where we wish to concentrate our heat, we place a body of sufficiently high **resistance, and send** a sufficiently strong current through it, **we can make it so hot** that it will *glow and give light.* **This is the** principle of all electric lamps of whatever **kind.**

We have stated **on** page 106 that for **light to be economically** produced, it is necessary to raise the body producing the light to the highest possible temperature; or, in other **words,** to concentrate the heat in a solid of the smallest **possible** size or with the smallest possible cooling surface.

Now let us take a platinum spiral such that a given current **makes** it just red-hot. We **get** a very little light. If **we take another** spiral composed **of half the** length of wire, and whose wire has half the section, it **will have the same** resistance as the first, and the same current passing **through** it, will produce the same quantity **of** heat **in it,** and expend the same quantity **of** heat in the steam-engine. The platinum having very much smaller cooling surface will be raised **to a much** higher temperature, **and will** become white hot **and give a** brilliant light.

If, the resistance being still kept constant, the surface be further reduced, the temperature will **be still** further raised, **aud the same** amount of heat **will** produce a still more brilliant light. It appears then as if by making the wire still thinner we could produce as much light as we pleased **from** a given quantity of heat.

The practical reason why we **cannot do** this is that **platinum and** all other metals fuse at what, in electric lighting, is a comparatively **low** temperature. Platinum fuses at about $2000°$ C. Further, **if heated in air, all known** substances rapidly oxidize **and burn away.**

The problem, then, which has **had to** be solved in electric lighting has been to obtain a s**ubstance** having a resistance of convenient magnitude which can be **heated** by the current **and** which is either indestructible **by intense heat** or if slowly destroyed is capable of easy and continuous renewal.

99. Use of carbon in electric lighting. Carbon, either alone **or** in conjunction with heated **air,** satisfies **these** conditions in a great **measure.** It has never yet **been** fused, and though it slowly oxidizes when heated in air, yet its destruction can be either guarded against or compensated for as is done respectively in the two great systems of electric lighting now in use, i.e. the incandescent and the arc system.

In the "incandescent"* lamps of Swan, Edison, Maxim, Lane-Fox, &c., the resistance, used to convert the current into heat, is that of a very fine thread or wire of carbon which is brought to a state of intense incandescence by the passage of a current through it, and which is protected from oxidation by being hermetically enclosed in a glass globe from which all the air has been exhausted. These lamps last for many months, if not too much heated.†

In the "arc" ‡ lamps the current is sent through two stout rods of carbon which touch each other end to end. As soon as the current is established the rods are separated a little way, and the current continues through the heated air, which is a partial conductor of high resistance. Great heat is produced, the "poles" of the carbon rods glow with an intense whiteness, and small particles of carbon becoming detached are heated in the air between, and form a luminous "arc" from one pole to the other, which adds to the light.

In this class of lamps the carbons being thicker can be raised to a much higher temperature than the carbon threads of "incandescent" lamps, and consequently they give much more light for a given quantity of heat, and are so much the "more efficient."

The carbon rods slowly consume away, and therefore have to be fed forward by suitable machinery. In arc lamps the expense of the carbon rods has to be added to the cost of producing the current in estimating the total cost of the light. Presently we shall describe various lamps, both "incandescent" and "arc," now in use. Lamps have to be constructed to use currents of certain strengths, and as each

---

* See page 115.

† If overheated even in a perfect vacuum the filaments are destroyed with more or less rapidity by some process analogous to mechanical disintegration—they are, as it were, shaken to pieces.

‡ See page 121.

I

lamp is intended to convert a certain definite quantity of electric energy into heat, we see how necessary it is for electrical engineers to comprehend the methods by which electrical quantities are measured and the standards to which they are referred, which we have already given an account of in Chapters VIII., IX., and X.

Questions on Chapter XV.

1. Show the analogies between a flow of water and a flow of electricity when used to convey heat from one place to another.
2. What is Ampère's theory of magnetism?
3. What is the principle of an electric lamp?
4. Why is carbon used in electric lighting?
5. What is an incandescent lamp?
6. What is an arc lamp?

CHAPTER XVI.

INCANDESCENT LAMPS.

100. Incandescent lamps. The class of lamps known as "incandescent" consists of a thin filament or wire of carbon enclosed in a glass globe from which the air has been exhausted.

On a suitable current of electricity being sent through the filament it becomes white-hot, or incandescent, and gives a light of from 1 to 100 candles according to its surface, and for a given surface according to the temperature to which it is raised. For a given temperature the durability of the filament depends on its uniformity, and on the completeness with which the air has been exhausted. Below a certain temperature, nearly corresponding to that of melting platinum, a well-made filament in a good vacuum is very durable. Under these conditions, lamps last six or twelve months of ordinary domestic work.

The chief incandescent lamps now in actual use are the Swan and Edison. They differ from each other in the methods of preparing the carbon filaments and in other details.

I propose to give a general description of the Swan lamp, which may be considered as a typical one, and to describe some of the processes used in its manufacture.

101. First exhibition of Swan lamps. The first public exhibition of incandescent lamps that was made in this country was made by Mr. Swan before the Society of Telegraph Engineers on November 24, 1880. The first exhibition in America was made by Mr. Edison.

102. Incandescent lamps. The carbon. In all incandescent lamps the filament consists of a thread of some vegetable substance which has been *carbonized* by heat.

103. The terminals. The ends of the filaments are connected to two platinum wires, which pass through the glass and are melted on to it. Platinum is used as, its expansion rate being about the same as that of glass, the latter does not crack in cooling.

104. The exhaustion. The life of the lamp depends in a large measure on the goodness of the vacuum. In order to get a good vacuum, various modifications of the Sprengel mercury-pump have been made, all having for their object to fit an instrument hitherto only used in laboratories to the more rapid processes of the factory.

105. Hot exhaustion. It was soon found that however perfectly the lamp was exhausted when cold, yet that the first time a current was sent through it a quantity of gas was driven out of the carbon itself, which injured the vacuum and caused the speedy destruction of the filament.

To surmount this difficulty the following plan was adopted, first, I believe, by Swan, and is now used by all makers of incandescent lamps. While the lamp is still attached to the pump a current of electricity is sent through the filament, sufficient to raise it to a somewhat higher degree of incandescence than will be used in actual work. All the gas driven out of the carbon is at once removed by the pump, and the lamp is sealed while the current is still passing.

106. Current, electric pressure, and copper. The horse-power expended in a lamp depends on the product of the current into the electromotive force at which it is supplied, and is therefore the same as long as the product is constant, whether the pressure is small and the current large or *vice-versâ*. We note that the higher the resistance of the filament, i.e. the longer and thinner it is, the more pressure is required to drive the current through it, and the less current is required to produce a given quantity of energy in the form of heat and light in the lamp.

The quantity of copper required for a conductor of given length depends only on the current it has to carry, and not on its pressure. *We thus see* that every improvement in the lamps which **enables the** filament **to be** made thinner and longer, and so diminishes **the** current **used,** proportionably diminishes the quantity of copper in **the mains.** As this copper **is one** of the most expensive items in an electric light plant, improvement in this direction is extremely important.

107. The Swan lamp. Since the introduction of the original Swan lamp, several modifications have been made in its form and other details. In the latest form the filament is much longer and thinner than in the old pattern, being about five inches long and ·005 inch in diameter. It requires a pressure of 100 to 120 volts to bring it to normal incandescence of 20 candle-power.

108. Process of manufacture. Figs. 56 to 61 show the manner in which the several parts of the Swan lamp are put together.

The filament is attached to its platinum wires and mounted on a glass bridge, as in fig. 56, little beads of glass being also formed on the wires where they are to pass through the walls of the lamp.

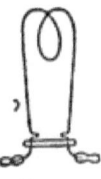

Fig. 56.

The globe is blown as in fig. 57, and with a sharp file is cut into two pieces, as in fig. 58.

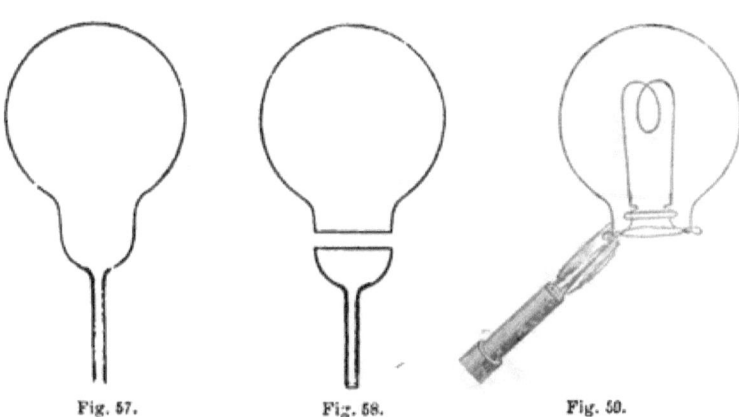

Fig. 57.   Fig. 58.   Fig. 59.

118   *School Electricity.*

The carbon and platinum wires are inserted, and the latter fused on by the blow-pipe, as in fig. 59.

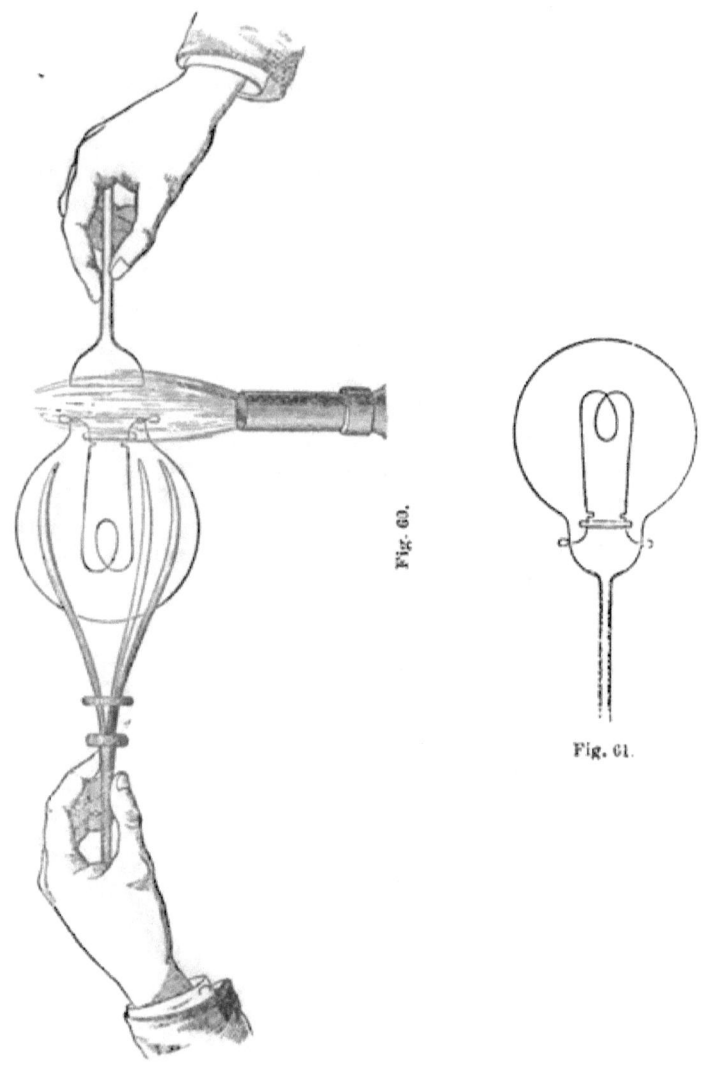

Fig. 60.

Fig. 61.

The two portions of the globe are joined again by the blow-pipe, **as** in fig. 60, and the lamp is completed in the form shown in fig. 61, and is ready to attach **to** the pump.

The following are the results of some experiments in the efficiency of the new Swan lamp :—

Candle-Power of Lamp.	Current in Ampères.	E.M.F. in Volts.	Resist. in Ohms. (Hot.)	Volt-ampères per Candle.	Candles per H.P.	Lamps per H P.
16	·62	98	158	3·74	199	12
16	·63	97·1	154	3·69	202	12
16	·726	78	107	3·5	213	12
18	·63	100	158·7	3·5	213	12
18	·64	99·1	154·8	3·52	212	12
18	·75	80	107	3·3	226	12
20	·65	102	157	3·32	225	11
20	·66	100	151·5	3·3	226	11
20	·76	82	108	3·1	240	12

109. Efficiency of incandescent lamps. The efficiency of any incandescent lamp can be altered at will. For instance, if we have a lamp intended to work at 100 volts and giving 200 candles per H.P. at that pressure, we shall find if we increase the pressure to 110 volts that we greatly increase the efficiency. It may be to 500 candles per H.P. or even more.

The higher the **efficiency** of the lamp the less coals are required to work **the engine** driving the dynamo in order to produce a given **quantity of** light.

But—the higher the **efficiency of the** lamp the less time it will last before breaking, i.e. **the greater** will be the annual cost for renewals of lamps.

We see therefore that where coal is cheap, or where we can get water-power to drive our dynamos, we ought to work our lamps at a low efficiency, but where coal is dear we should work them at a high one.

Inventors of new lamps generally state that their lamps are thirty per cent. higher efficiency than those in common use, and invite tests, which being made prove that their statement is correct. This high efficiency is however obtained at the expense of durability of "life," but the diminution of life is not detected during the short time that the exhibition of the lamps lasts.

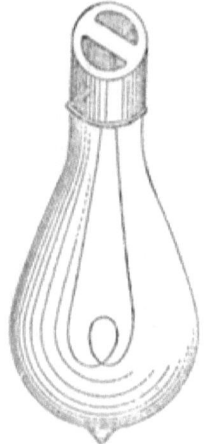

Fig. 62.

110. The holder. Swan lamps are more generally made with a cap, as shown in fig. 62, and the two insulated semi-circular plates of metal are connected to the terminals. This cap or neck fits into a socket in the chandelier provided with a bayonet joint, and two springs on it press on the plates of the lamps. These springs are connected respectively to the mains, and serve the double purpose of making contact and keeping the lamp in its place.

QUESTIONS ON CHAPTER XVI.

1. Describe, with sketches, the process of manufacture of a Swan lamp.

2. When should Swan lamps be worked at high efficiency, when at low?

3. What is the average efficiency at which Swan lamps are worked?

CHAPTER XVII.

ARC LAMPS.

111. Arc lamps. In arc lamps, as we have already stated,* the resistance which converts the current into heat is that of the heated air between the ends of two carbon rods, from one to the other of which the current passes. The light is produced by the incandescence of the end of the carbon poles and of the minute particles of carbon which become detached and float in the heated air between them. The heated air containing the particles of carbon forms what is called the "electric arc."

The carbon rods vary in diameter from $\frac{1}{8}$ inch in the smallest lamps made, to $3\frac{1}{2}$ inches in a lamp recently constructed by the Brush Company.

The carbon rods slowly burn away, and therefore have to be continuously fed forward by suitable machinery, so as to keep "the resistance of the arc" as constant as possible. On the steadiness of the feeding machinery, and on its sensitiveness to minute changes in the resistance, depend in a great measure the steadiness and freedom from flickering of the light.

It is also necessary that all arc lamps should light themselves when the current is started, i.e. that when no current is passing, the carbons should be in contact, and that when the current commences to flow, they should be instantly separated to a distance giving an arc of the required resistance. This distance will vary according to pressure, size of lamp, &c., from $\frac{1}{10}$ inch to $\frac{3}{4}$ inch.

An immense number of regulators have been constructed

* Page 113.

by different inventors, but they may all be divided into some three or four general types. I propose in the present chapter to describe two lamps only, namely, the Serrin and the Crompton. I have selected these two, as the Serrin is very easy to understand, and the Crompton, though more complicated, is about the best lamp at present in practical use.

112. The Serrin lamp is only used for lighthouses.

The qualities required in lamps are different according to the service for which they are intended.

For lighthouse work it is absolutely necessary that the light shall never be extinguished for an instant, that the mechanism shall be strong, and that an ordinary lightkeeper shall be able to manage it. Expense, weight, and bulk are matters of no consideration whatever; neither is there any objection to the mechanism being below the arc, as the light is not required to be directed vertically downwards. Slight pulsations in the light are not a serious defect. The arc must always be kept in the focus of the reflector, so both carbons must be fed forward.

In lamps intended for street lighting the chief consideration is steadiness and freedom from flickering. They must be moderately cheap, and not too heavy to hang on an ordinary lamp-post. The whole of the mechanism must be above the light, so that shadows may not be cast downwards.

A temporary extinction of the light, though much to be deprecated, would not, as in the case of the lighthouse lamps, be likely to have consequences fatal to life, and therefore strength of machinery need not be studied to the exclusion of all considerations of economy.

Further, street lamps must be so constructed that several can be worked off one machine, and so that the accidental extinction of one shall not affect the rest. As a slight lowering of the position of the light is not objectionable, one carbon may be fixed and only one fed forward.

Arc lamps are not suited for the interior illumination of rooms, but when so used considerations of perfect freedom from flickering outweigh all others. In this case the lamp must be also so adjusted as to be free from the hissing sound which is often produced by an electric arc.

All lamps should be constructed so that they will burn from dark to daylight of a winter night, say sixteen hours, without attention or requiring new carbons.

When lamps are worked by a direct current, the positive carbon consumes away about twice as fast as the negative; with an alternating current the consumptions are of course equal. The adjustments of springs, &c., required when direct and alternating currents are used, are somewhat different.

113. The Serrin lamp. Fig. 63 is a drawing of a Serrin lamp.

The base of the lamp contains clockwork, which is actuated by the weight of the upper carbon. The racks carrying the upper and lower carbons are connected by cog-wheels, so that as the upper carbon sinks the lower one rises to meet it. When the lamp is to be used with alternating currents the cog-wheels gearing into the two racks are of the same size, and the carbons advance equally. When direct currents are to be used, the cog-wheels are so proportioned that the + carbon moves twice as fast as the − one. As the carbons move, the star-wheel, which is the last wheel of the train, revolves very rapidly, and a very slight brake applied to it is sufficient to lock the carbons. When the carbons are in contact a current can be sent through the lamp. The current passes through the electro-magnet, which attracts its armature, and pulling a lever draws down the lower carbon-holder, and, separating the carbons, forms the arc. At the same time the lever locks the star-wheel, and prevents the carbons from moving.

As the carbons burn away the arc gets longer, and as its resistance increases, the current in the magnet gets weaker, and the armature is drawn a little way from it by the spring. This releases the star-wheel, and the carbons approach each other till the current has recovered its proper strength, when the armature is again attracted and the carbons locked. This adjustment is repeated at intervals until the carbons are consumed.

The upper carbon can be brought exactly into line with the lower one by means of the screws seen at the top of

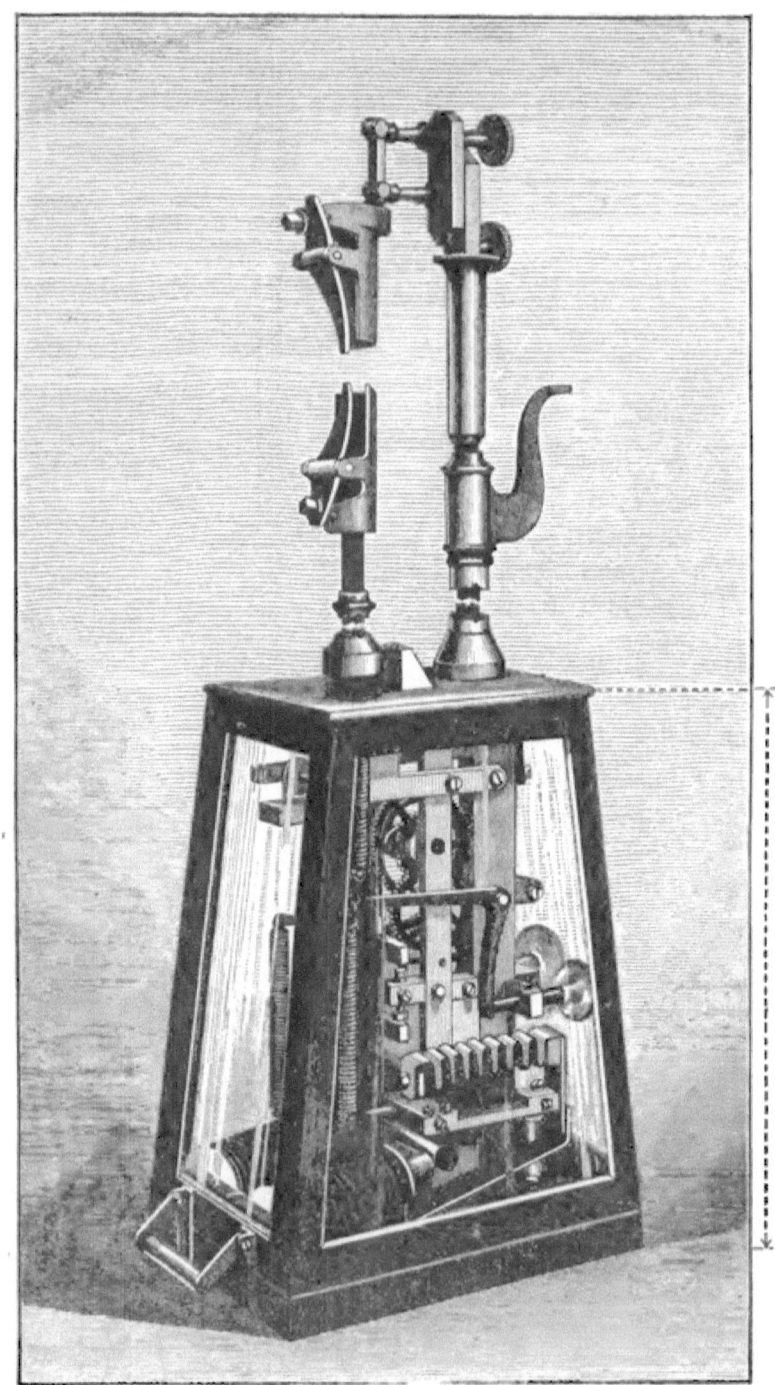

Fig. 63.

fig. 63. Of the two screws seen on the left-hand side of the clock-case, one adjusts the tension of the spring and so regulates the length of the arc, the other enables both carbons to be raised or lowered together without altering their distance apart, so as to place the arc exactly in the focus of the reflector. The position of the arc should be level with the top of the little bracket which is pivotted on the tube inside which the upper carbon-rack slides. The bracket can be turned round so as to be close to the arc, or can be turned back out of the way. The lamp is simply slid into position on two brass rails, to which the wires from the machine are attached. In case of any accident to the lamp it can be removed and a spare one substituted in a few seconds, as placing the lamp in position at once makes the connections.

114. The Crompton lamp. The latest type of lamp made by Messrs. Crompton and Co. is known as the D.D. (double differential) lamp. It is the joint invention of Messrs. Crompton and Crabb, and shows a marked improvement on the older forms in point of simplicity of construction and also in regulation. This latter is effected by a brake-wheel driven by the rack rod attached to the upper carbons.

Referring to fig. 64, B and B_1 are the rack rods carrying the positive carbons. Sliding on each of these is a light gun-metal sleeve, S S_1, carrying spindles, to which are attached the two large brake-wheels E E_1, and between them the pinion which gears into the racks. To each side rod is pivotted a broad lever L L, at the other end of which a chain is fastened, connecting it to the hollow core of the solenoid vertically above. This solenoid is what is called a "differential coil," i.e. it is wound with two sets of wires M and G. The action of a current in M is to draw the iron core upwards, that of one in G to pull it down. The regulating gear, which we shall describe immediately, is so constructed that when the core is raised the carbons are separated, when it is lowered they are brought nearer together.

The coil M, which separates the carbons, is wound with thick wire, and is "in series" with the arc, and therefore if the current through the arc gets too strong the increased

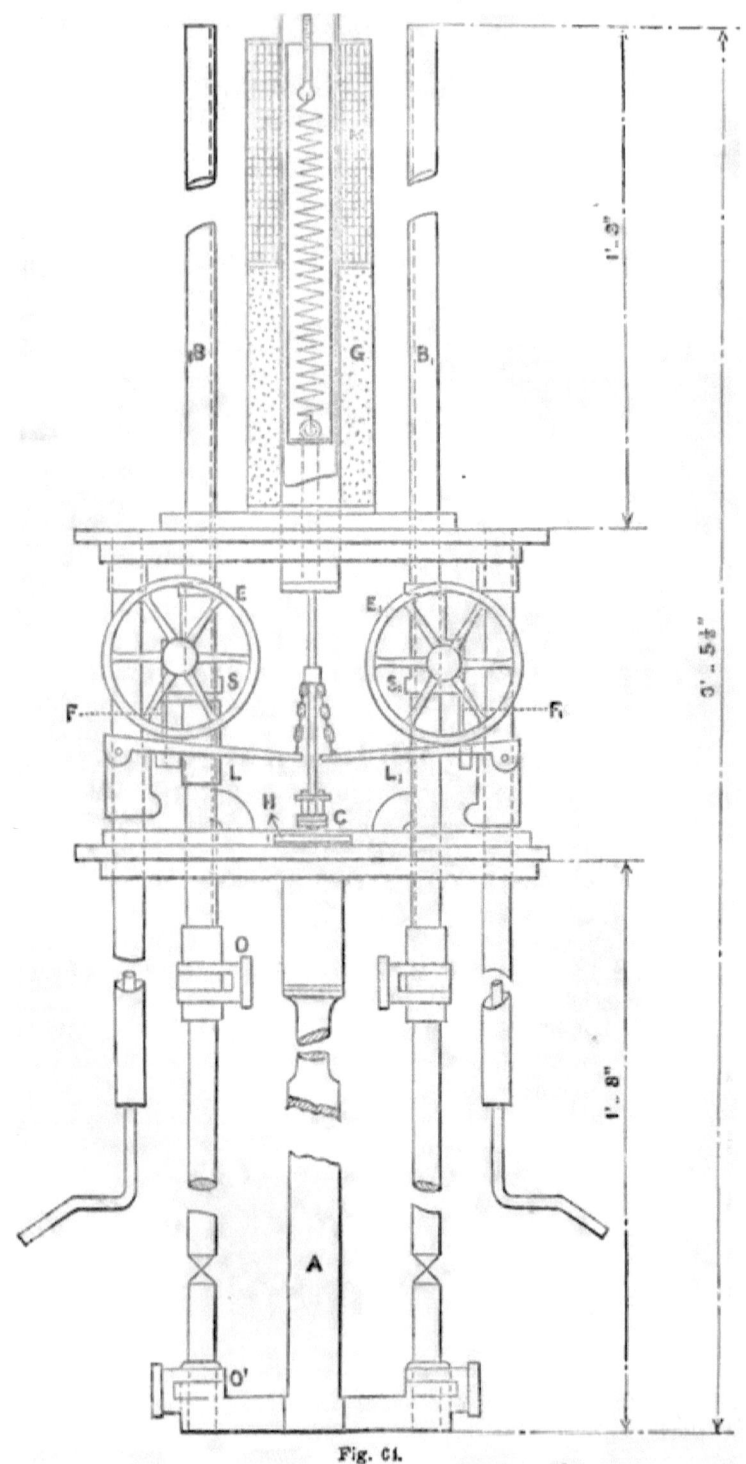

Fig. 64.

strength of **the same** current in **the coil** M so actuates **the** machinery that it lengthens the **arc,** and so, increasing **its** resistance, decreases the **current.**

The other coil G is **wound with** fine wire and is "in **shunt"** with the arc, that is, part of the current goes through the arc and part through the shunt. If the arc gets too long, its resistance increases and a greater portion of current goes through the shunt coil; this pulls the core downwards and shortens the arc.

The following is the machinery by means of which the motions of the iron core move the carbons. In reading the description constant reference must **be** made to the diagram, fig 64, when it will be found **not very** difficult **to** understand.

Projecting vertically downwards from each sleeve **to a** distance from the centre of the spindle about equal to the radius of the brake-wheels, is **a** stout pin **or** finger, F F$_1$, the use of which **we** will try **to** make clear. Suppose the **rack** rod to be drawn up; then, if the lever be pulled by the solenoid above the horizontal position, the whole weight of the rod and carbon is supported on the edges of the two brake-wheels, and the friction of them on the surface of the levers is sufficient **to** prevent **their** revolution; hence this rack rod cannot run down; **but if the** levers **be below** the horizontal, then the **weight is carried by the** finger projecting from the sleeve, **as shown at F,** the wheels are free **to** turn, **the** rack runs down, **and** continues **to do so** until the positive and negative carbon points come in contact. Now let the current be switched on: by its passage through the main **wire** of the solenoid the levers are raised, striking **the** arc, and at the same time applying the brake to the wheels. The shunt current then flows, and the arc takes its proper length. If **this** becomes too great, **the** increased current through the **shunt draws** down the core and levers; the brake-wheels are **left free to** revolve and **the** arc shortens. On the other hand, **if the carbon** points **be** too close, the levers are raised, bringing with them the rack rods and upper carbons.

Making the finger projecting from one sleeve longer than

that from the other, determines which pair of carbons shall begin to burn first, because, on switching on, that pair which has the longer pin will be the last to break contact, and will therefore originate an arc in so doing. It will be easily seen from fig. 64, that on the core being raised the lever L_I will apply the brake before the lever L does, hence it may be said that the rack rod B_I gets a start on B; its carbon points are separated before those of B, and are kept a greater distance apart until the latter are consumed. When this is the case, the rack rod B_I is prevented from further fall by a stop and can no longer feed, hence the arc will lengthen, the shunt current will increase, and the other rod B_I, which can still feed, will be allowed to descend until its carbons touch, starting a fresh arc. The core is raised again, the fresh arc burning instead of the old one, and everything goes on as before until the second pair of carbons is consumed.

115. Carbons for arc lamps. Davy's first experiment with the electric arc was carried out with pieces of wood charcoal as electrodes, but it was at once seen that electrodes formed of such a soft material could not be of much practical use, as they burned away too rapidly, and gave off coruscations of dangerous sparks. It may be here mentioned that, seventy years later, Gaudoin, at Paris, again reverted to the use of rods of wood charcoal, the density of which he increased to any required extent by filling up the pores with various hydro-carbons, alternately soaking the rods in liquid hydro-carbon, and firing them until they gave a metallic ring. But for many years the carbon electrodes used for all experiments with the electric light were strips sawn from the graphitic deposits found in gas retorts.

In the carbon electrodes as now used coke or graphite is finely powdered and washed in alkaline cells to get rid of the silica and earthly impurities, after which it is ground in a pug-mill, with sufficient syrupy or tarry hydro-carbon to agglutinate it into a stiff paste. This paste is then pressed into rods of the required form by being forced through moulds or dies.

Sometimes the pressure is applied endways, the paste

being forced out in a continuous rod of the proper thickness, and cut off into lengths as required. Another plan is to apply the pressure sideways, the dies being divided longitudinally into two parts. In the latter case, several rods are usually pressed at one time. The rods thus formed are carefully dried, and afterwards fired in kilns, having been previously packed in air-tight boxes, and embedded in coke dust. After one firing they are generally found to be porous, and require soaking in syrup and a second time firing. Some of the makers repeat this process more than once.

The following points are aimed at in the production of a perfect carbon for arc lighting :—

1. Freedom from all matter other than carbon.
2. Regularity of density.
3. Mechanical perfection of form.
4. Low electrical resistance.

116. Purity. It is not sufficient that the coke or other powdered carbon from which the rods are made should be in the first instance free from earthy or metallic impurities, but a great deal depends on the care taken in subsequent processes of manufacture to insure that the volatile gases, most of them hydro-carbons, are thoroughly expelled during the process of firing. The reasons for this are as follows :—

Carbon being the most refractory of any known substance, disintegrates to pass across the arc at the highest possible temperature, and as the whiteness of the colour, as well as the amount of the light given by the electric arc, depends entirely upon the temperature of the arc,* it is evident that any admixture of substance other than carbon will, by lowering the mean temperature at which the electrode is disintegrated, tend to diminish the amount of light given.

Observations taken with the spectroscope show that at the times when the arc distils itself free from all foreign salts, the colour of the light is most intensely white, and the amount of light is at a maximum; and it is almost certain the temperature is also at a maximum. At the same time it is observable that the conductivity of the stream of matter passing across from one electrode to another is at a minimum;

* Compare p. 112.

hence if the difference of pressure at the two sides of the arc is maintained constant, the arc will be shortest at the times that the carbon is purest, and the action being then confined to the smallest possible space, will be intensified in proportion, resulting in greatly increased light and economy. The addition of the smallest portion of any material which volatilizes at a lower temperature than carbon itself, increases the length of the arc with the results the reverse of those above mentioned.

Gaseous impurities, the chief offenders being hydrocarbons, are peculiarly annoying in this respect. When they are present to any extent, they always break out at irregular intervals as blowers of gas, which are comparatively of high conductivity. These jets play around the crater proper in a most irregular manner, and are the chief cause of the flickering so often complained of in arc lighting. It is quite common to see the arc start from a point very high up on the cone outside the crater, and forming a curved and rapidly-moving jet towards a point on the cone of the negative carbon. At such moments the light is found diminished to one-third or one-quarter of its normal brilliancy, and as the lamp adjusts itself to the new conditions by automatically lengthening the arc, a vibratory or reciprocating action is set up in the lamp, which greatly intensifies the mischief.

Silica and other earthy matter when present in carbon, in addition to lowering the arc temperature, form a more or less bulky ash, which falls down into the surrounding globes, and is extremely unsightly. When continuous currents are used, this ash accumulates on the negative electrodes, and thus produces ugly and otherwise objectionable shadows.

For the method of satisfying the three latter requirements of a good carbon the reader is referred to my "Electric Lighting," p. 103.

117. The size of carbon electrodes. The economical efficiency on the one hand, and the steadiness of the light on the other, depend very greatly on the diameter of the carbon electrode employed with a given current. Within certain limits the one is in inverse ratio to the other. That is to say, as we decrease the diameter of the carbon, we increase the amount of light from a given current, and decrease its steadiness. The diameters most commonly used have been as follows:—

For currents from 7—12 ampères	9 m.m. to 11 *m.m. diam.
,, 12—18 ,,	11 m.m. ,, 13 m.m. ,,
,, 18—25 ,,	13 m.m. ,, 15 m.m. ,,
,, 25—40 ,,	15 m.m. ,, 18 m.m. ,,
,, 40 upwards	18 m.m. ,, 20 m.m. ,,

It is extremely difficult to obtain carbons above 20 m.m. of sufficient homogeneous texture to give a steady light with lamps having automatic regulation. All carbons above this diameter are used with the large arcs as search lights for military and naval purposes, and most commonly worked by non-automatic hand regulators, and as all the best modern projectors are fitted with an arrangement for viewing the arc from the side, the attendant in charge can quickly set right any irregularity caused by change of density.

If the above rules as to diameter proportionate to current are departed from, the following results take place :—

If a carbon of too small diameter is used, the positive carbon cores away so rapidly as to cut down the sides of the crater, and the intensely-heated portion of the carbon extends outside the walls of the crater proper; in this way a great amount of the light which ordinarily would be thrown downwards on to the surface required to be illuminated, will be wasted in the upper part of the spherical angle. In addition to this loss, there is great danger of the carbon becoming red-hot from end to end, and thus wasting away like a candle in a hot oven. Again, the light also becomes very unsteady; the arc does not pass steadily between the points of the cone and the crater proper, and the lamp burns for a shorter time than it ought to do.

On the other hand, if carbons of too great diameter are used, they do not point properly, and consequently, the angle of the lower carbon being very obtuse, throws down a large shadow. A great part of the electric energy, instead of being utilized in producing an intense temperature at

* One m.m. (millimetre) is practically equal to $\frac{1}{25}$ inch.

the crater, is wasted in heating the external portion of the massive carbon to red heat.

Nevertheless there is no doubt that the use of large carbons has greatly extended during the last few years. Although energy is wasted, and consequently less light is afforded by a given current, there is a great increase in steadiness, and the lamps burning for longer hours do not require so much attention.

The following table gives the result of experiments on the illuminating power per electrical H.P. of carbons of varying diameters :—

Ampères.	18 m.m.		13 m.m.		11 m.m.	
	Candles.	Volts.	Candles.	Volts.	Candles.	Volts.
5 to 14	2660	60·0	2278	48·0	2549	52·2
14 ,, 18	4400	39·0	3514	46·9	—	—
18 ,, 25	5263	42·75	3637	42·9	—	—

QUESTIONS ON CHAPTER XVII.

1. Describe the Serrin arc lamp.
2. Describe the Crompton arc lamp.
3. What are the principal requirements for good carbons?
4. Why are volatile impurities in carbons objectionable?

CHAPTER XVIII.

DYNAMO MACHINES.

ALL dynamo machines consist of apparatus for moving coils of wire past magnets or magnets past coils of wire. The result of so doing is to produce electric currents in the wire of the coils. These currents are led out from the machines and utilized.

Dynamo machines may be divided into two great classes, *direct* and *alternating*.

118. Direct and alternating currents. A direct current is a current which always flows in the same direction, an alternating current is one whose direction is constantly being reversed. The alternating currents used in practical electric lighting are reversed about 5000 times per minute. Each class of current has its special uses and advantages.

119. Alternating current machines. Alternating current machines consist generally either of two fixed wheels with magnets all round their rims, and of a wheel with coils round its rim which revolves between them, or else of one revolving wheel carrying magnets revolving between two fixed wheels carrying coils.

The largest alternating machines which have been constructed, are those which I have lately erected at Greenwich and at Paddington. The following is a description of the Greenwich machine. Fig. 65.

It consists of a wrought-iron wheel, carrying the magnets, which is eight feet diameter at the magnet centres (8 ft. 9 in. over all). The magnets, which are thirty-two in number, consist each of a cylindrical core of soft wrought-iron

which passes right through the wrought-iron disc, and projects equally on both sides of it.

Brass bobbins, containing the magnet wire, are slid on to the projecting portions of the cores, and are kept in their places by the pole plates, which are afterwards attached. The revolving wheel is built up of sheets of boiler-plate riveted together, and strengthened by two cones of boiler-plate, placed one on each side. The cones and disc are separated by cast-iron distance-pieces, as shown in section in fig. 65.

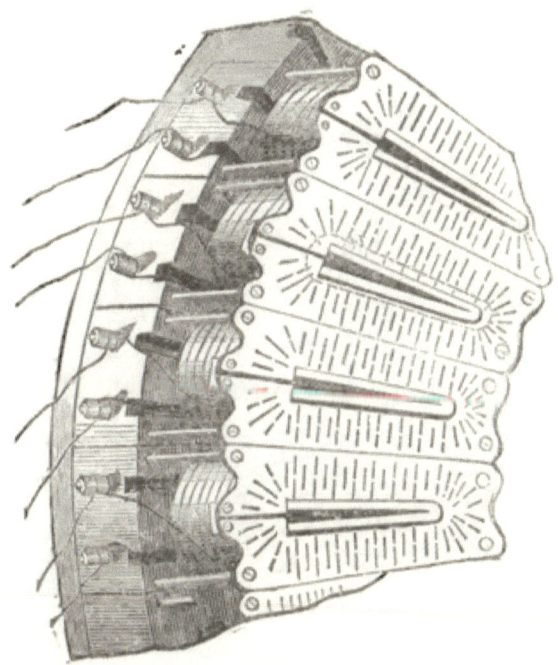

Fig. 61.

This wheel revolves between two fixed iron rings carrying the armature coils.

The fixed coils are secured to cast-iron frames, but the cores are prolonged, so that the frames are set back into a field of weak magnetism. Fig. 66 shows some of the fixed coils. Their flanges are made of German silver to check the circulation of currents in them.

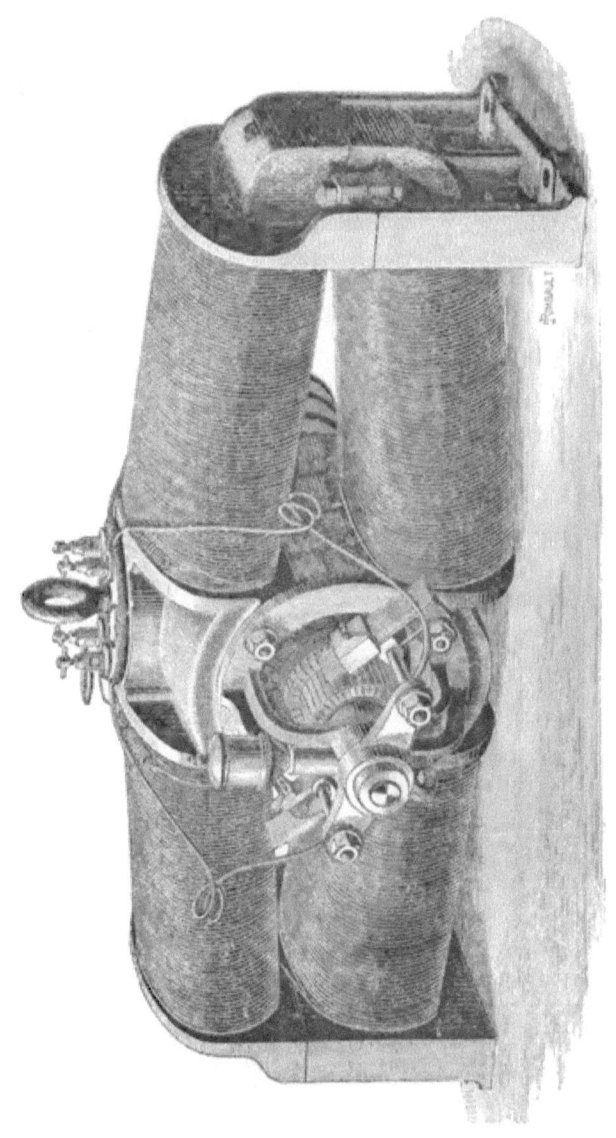

Fig. 67.

Direct Current Machines.

The eight-foot machine runs at 140 to 180 revolutions per minute, and so is connected direct to the steam-engine without bolting. The total weight is twenty-two tons, and the total weight of the revolving magnet wheel seven tons. The only rubbing contact is that where the exciting current produced by a direct current machine enters the revolving magnets.

As the magnet poles pass the soft iron poles of the fixed coils they magnetize them by induction, and as consecutive magnet poles are of opposite polarity the magnetism of each of the iron coils is reversed 32 times each revolution, or $32 \times 140 = 4480$ times per minute.

At each reversal a current is induced in the wire of the coils which flows out into the mains, the direction of the current being therefore reversed 4480 times per minute.

120. Direct current machines. The direct current type of machine may be divided into two sub-types, namely, the "Gramme" type and the "Siemens" type.

Fig. 67 is a view of a machine of the Gramme type, known as the Burgin machine.

It consists of a pair of large magnets, the poles of which are the curved masses of iron seen at the top and bottom of the machine. One, say the top one, is a N. pole, the other a S. pole.

The armature revolves in the cylindrical opening between the poles. The method of its action will be best understood by a diagram. Fig. 68.

Fig. 68 is a diagram of a machine of the Gramme subtype. The armature consists of a ring of soft iron round which wire is wound as a continuous spiral, forming a closed circuit.

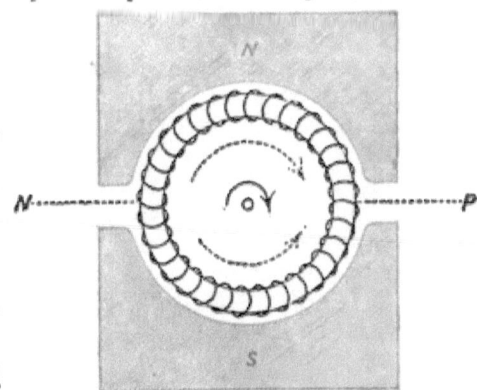

Fig. 68.

It revolves between two poles of opposite names, the lines of force from which terminate in the ring.

As the ring revolves these lines of force are cut by the moving wires, and electro-motive forces are generated in the two halves of the ring *in opposite directions,* so that they meet and oppose one another at the neutral points N P, as in fig. 69.

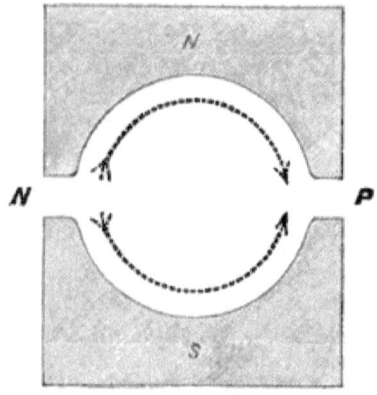

Fig. 69.

As long as no further connections are made, no current is generated, and no H.P. expended. If, however, the points N P are connected through an external circuit, such as a number of lamps, the two halves of the ring will act like two batteries in parallel circuit, and a current will flow, as in fig. 70.

We see that, owing to the ring being in motion and the neutral point necessarily at rest, a permanent connection between the line and the wire in the ring cannot be made, but a special device has to be employed.

121. The collector. The collector is made of a barrel of wood or other insulating material, shown in the centre of fig. 71, on which are a number of insulated metal strips. Each of these strips is connected by a wire to the part of the spiral wire opposite to it. Two metal brushes press or rub on the strips at the points P N, where the opposite electro-

Theory of the Gramme Ring. 139

motive forces diverge and join again. These brushes convey the current to the external circuit.

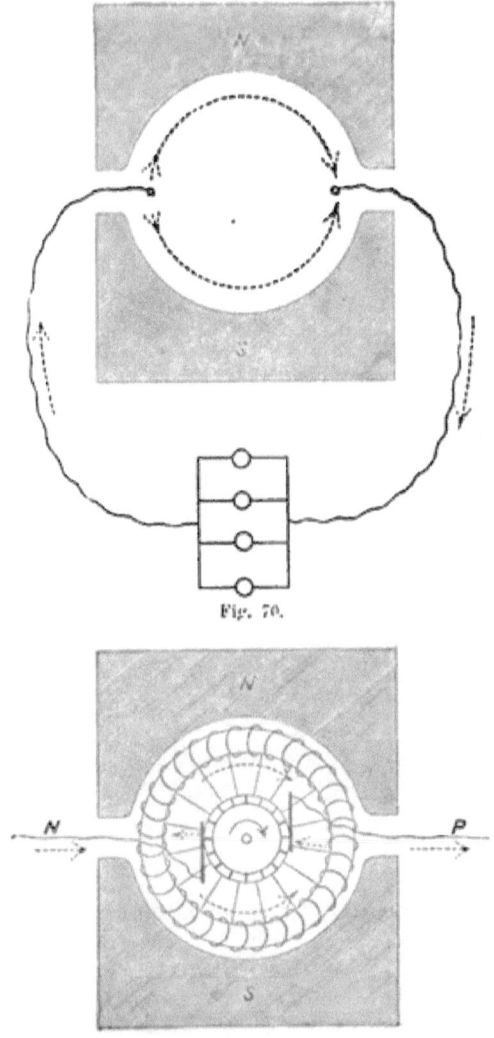

Fig. 70.

Fig. 71.

140   *School Electricity.*

122. The Siemens sub-type. In this type of machine the wire is wound longitudinally round an iron barrel. It differs from the Gramme ring by the omission of those parts of the wire which pass inside the ring. Fig. 72 shows the Gramme type in section, fig. 74 the Siemens type, and fig. 73 an imaginary intermediate type.

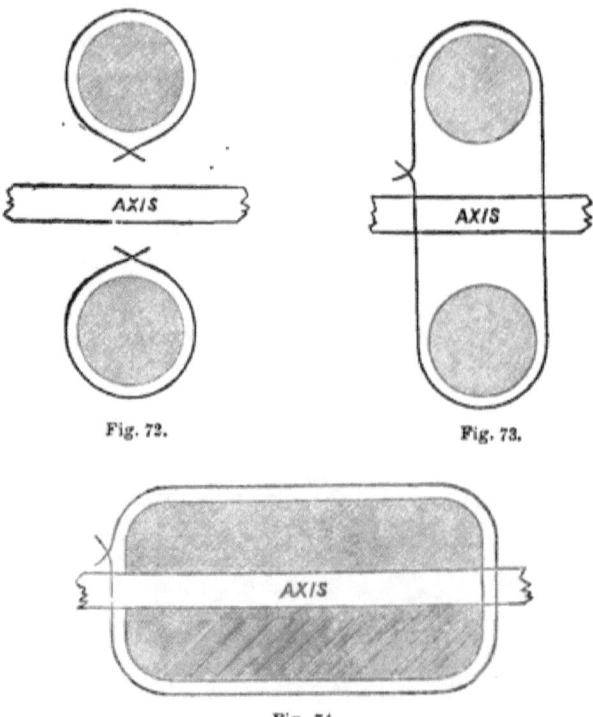

Fig. 72.   Fig. 73.

Fig. 74.

The collector in this type of machine is similar to that in the Gramme type.

123. Production of magnetism in the field magnets.—The direct type of machine can "excite" its own magnets, as a portion of the current generated is sent round their coils. The alternating type has to have its magnets excited by a small auxiliary direct-current machine. There are three forms in which the direct-current machines are constructed to excite their own magnets, and they are called respectively, "Series-wound," Shunt-wound," and "Compound."

In the *series-wound* dynamos the magnets are wound with a comparatively short length of wire thick enough to carry the whole current generated, and are connected *in series* with the armature and brushes. The first time such a machine is worked it must be excited separately by a battery or by another machine.

When once the magnets have been excited, a feeble *residual magnetism* will remain in them. On the machine being worked this feeble magnetism develops a feeble current in the armature ring. This feeble current passing through the magnets strengthens them, and they in turn strengthen the current in the armature. These alternate reactions go on until the maximum current that the machine can give is being generated.

This maximum is limited first by the external resistance, or, if the machine be short-circuited, it is limited by the magnets approaching their saturation-point, and by the internal resistance of the armature.

The *practical* current that can be taken out of such a machine is limited by the capacity of the wire to carry it without undue heating.

Short-circuiting a series-wound dynamo will do either one of three things: burn through the insulator, or by the extra H.P. absorbed, throw off the belt or pull up the steam-engine.

In *shunt-wound* dynamos the magnets are wound with a large quantity of thin wire, which is connected to the armature brushes *in quantity* with the lamps or other external circuit. The currents in the magnets and lamps then divide according to the ordinary rules of divided circuits. The same alternate reinforcement of the current and magnets goes on as in the series machines. Short-circuiting a shunt-wound dynamo simply stops the current, as it removes all the current from the magnets.

In *compound* machines the magnets are wound partly shunt and partly series.

Questions on Chapter XVIII.

1. Describe the Gordon dynamo machine.
2. Explain the theory of the Gramme ring.
3. Explain the Gramme collector.
4. Describe the Burgin machine.
5. Describe the Siemens direct current machine.
6. What is series winding?
7. What is shunt winding?

CHAPTER XIX.

ELECTROLYSIS.

124. Description of the phenomenon. If a body which is a compound of two substances is in a liquid state, and has a current of electricity passed through it, it is in general decomposed into its constituent elements, one of which appears at each of the points where the current enters and leaves the liquid.

If two platinum wires be immersed in acidulated water, and connected to a battery, the water will be decomposed; and hydrogen will appear at the negative pole, oxygen at the positive, and the volume of the hydrogen produced will always be double that of the oxygen.

If a solution, say of **sulphate** of copper, is substituted for the acidulated water, **copper is** deposited on the negative pole, while sulphuric **acid** is liberated at the **positive.**

125. Faraday's nomenclature.* The process of resolving compound bodies into their constituents is called *Electrolysis.* The bodies acted on are called *Electrolytes.* The poles at which the decomposition takes place are called *Electrodes.*

The electrode attached to the zinc of the battery is called the *cathode;* and the other, the *anode.*

The products of decomposition are called *Ions*, those which go to the anode are called *anions* and those which go to the cathode *cations.*

Thus chloride of lead is an *electrolyte,* and when *electrolysed,* by having the *electrodes* of a battery immersed in it, evolves

* "Exp. Res.," 665, vol. i. p. 197.

the two *ions*, chlorine and lead, the former being an *anion*, the latter a *cation*.

126. Laws of electrolysis.* *No elementary substance can be an electrolyte.*

For by definition, an elementary substance is that which cannot be separated into two constituents.

Electrolysis only occurs while the body is in the liquid state.

The free mobility of the particles is a necessary condition of electrolysis, for the process can only take place in one of two ways.

The molecule next one of the electrodes is decomposed. One constituent of it goes to the near electrode, and the other *either* travels to the other electrode *or* combines with a constituent of the molecule next to it, setting free a portion similar and equal to itself; which in its turn combines with the corresponding portion of the molecule next to it, and so on. In either case the free mobility of the particles is an essential condition.

Nevertheless, electrolysis sometimes occurs in viscous solids; but only in proportion to their fluidity.

Fused nitre is an excellent conductor in the liquid state. If, however, a cold platinum wire connected to a battery be dipped into it, electrolysis does not commence till the crust of solid nitre, which is formed round the cold wire, has had time to re-melt.

On this point Professor Maxwell † says,—

"Clausius,‡ who has bestowed much study on the theory of the molecular agitation of bodies, supposes that the molecules of all bodies are in a state of constant agitation, but that in solid bodies each molecule never passes beyond a certain distance from its original position, whereas, in fluids, a molecule, after moving a certain distance from its original position, is just as likely to move still farther from it as to move back again. Hence the molecules of a fluid apparently

---

* See Miller's "Chemistry," 4th ed. vol. i. p. 516.
† Maxwell's "Electricity," 256, vol. i. p. 309.
‡ Pogg. Ann. Bd. ci. S. 338 (1857).

at rest are **continually** changing their positions, and passing irregularly **from one part** of the fluid to another. In a compound fluid he supposes that not only the compound molecules travel about in this way, but that, in the collisions which occur between the compound molecules, the molecules of which they **are** composed are often separated and change partners, so that **the** same individual atom is at one time associated with **one** atom of the opposite kind, and at another time with another.

"This process Clausius supposes to go on in the liquid at all times; but when an electro-motive force acts on the liquid, the motions of the molecules, which before **were** indifferently in all directions, **are** now influenced by **the electro-motive** force, **so that** the positively charged molecules have **a** greater tendency towards the cathode than towards the anode, **and** the negatively charged molecules have a greater tendency to move **in** the opposite direction. Hence the molecules of the **cation** will, during their intervals **of** freedom, struggle **towards the cathode;** but will continually be checked in **their course by** pairing for a time with molecules of the anion **which are also** struggling through the crowd, but **in** the opposite direction."

The direction **of the** molecules **is always the** same with regard **to** the direction of the battery **current.**

The following very instructive experiment for showing the definite direction of electrolytic **force is** due **to the** late Dr. **W. A.** Miller. He says,—

"Let* four glasses be placed **side by** side as represented in **fig. 75,** each divided into **two** compartments by a partition **of card** or three or four folds of blotting paper, and let the **glasses be in** electrical communication with each other **by means of** platinum wires which terminate **in** strips of platinum **foil.** Place in the glass No. 1 a solution of potassic iodide mixed with starch; in No. **2, a** strong solution of common salt, coloured **blue with sulphate of** indigo; in 3, a solution of ammonium **sulphate, coloured blue** with **a** neutral infusion of the **red cabbage;** and **in 4, a solution of** cupric **sulphate.** Let the **plate H** be connected **with the** positive

* Miller's "Elem. Chem.," vol. i. p. 517.

L

wire, and let A complete the circuit through the negative wire. Under these circumstances iodine will speedily be

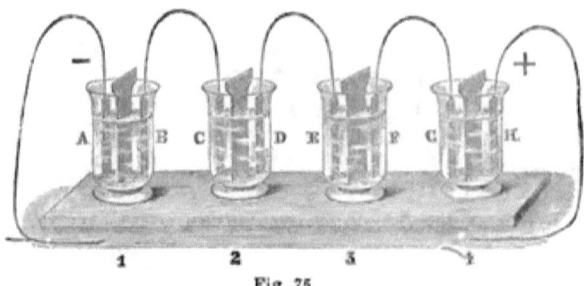

Fig. 75.

set free in B, and will form the blue iodide of starch; chlorine will show itself in D, and will bleach the blue liquid; sulphuric acid will be seen in F, and will redden the infusion of cabbage; sulphuric acid will also be liberated in H, as may be seen by introducing a piece of blue litmus paper, which will immediately be reddened; whilst a piece of turmeric paper will be turned brown in A, from liberated potash; in C, it will also be turned brown by the soda set free; in E, the blue infusion of cabbage will become green from the ammonia which is disengaged; and in G, metallic copper will be deposited on the platinum foil."

For a constant quantity of electricity, whatever the decomposing conductor may be, whether water, saline solutions, acids, fused bodies, &c., the amount of electro-chemical action is a constant quantity. That is, the same quantity of electricity will always produce the same amount of chemical effect.*

The same current electrolyses different quantities of different substances, but the proportion of one to the other depends *only* on their chemical equivalents.

Thus, if a current from a battery be sent through a series of troughs containing respectively,—

Water . . . . . . (H_2O),
Fused plumbic iodide . . . (PbI_2),
Fused stannous chloride . . . ($SnCl_2$),

---

* Faraday's " Exp. Res.," 505, vol. i. p. 145.

Electrolysis. 147

then for each 65 milligrammes **of** zinc dissolved in **any** one cell of the battery, there **will** be produced,—

$(2 \times 1) = 2$ milligrammes of hydrogen,
16 ,, of oxygen,
207 ,, of lead,
(2 × 127) $= 254$,, of iodine,
118 ,, of tin, and
$(2 \times 35.5) = 71$,, of chlorine;

and these numbers,—

65, 1, 16, 207, 127, 118, 35·5,

correspond **to** the chemical equivalents of the elements respectively.

If three similar vessels A, B, C, with platinum plates, **and** containing acidulated water, be arranged as in fig. 76, **and**

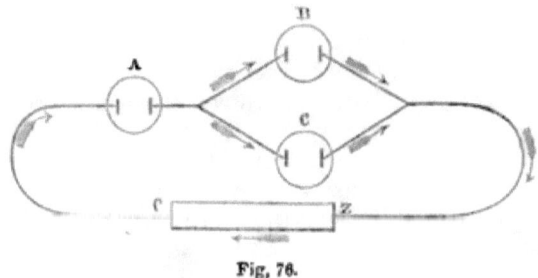

Fig. 76.

a battery current passed through them, the sum of the quantities of gas produced in B and C will be exactly equal to that produced in A.

127. The voltameter. This fact enabled Faraday to invent the *Voltameter*, which consists of a trough containing acidulated water, and having electrodes inserted in it. Receivers over the electrodes collect the gas produced. *The quantity of gas produced per minute is an absolute measure of the mean strength of the current* during that time; and the total quantity of gas **is a** measure of the total strength of the current.

It is necessary to collect the gases separately, as chemi-

L 2

cally clean platinum has the power of inducing their recombination.

Fig. 77 shows a common form of the instrument.

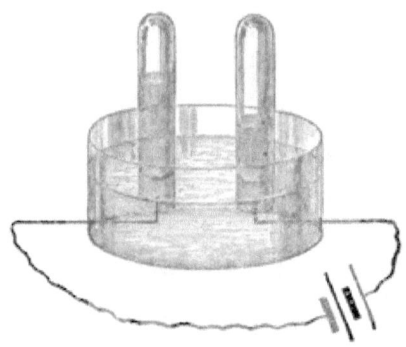

Fig. 77.

The tubes are previously filled with water and inverted over the electrodes. As the gas rises it displaces the water. The amount of gas formed is known by graduations on the tubes.

Electrolysis is of great practical importance, for nearly all the operations of plating, whether with copper, silver, or gold, are performed by making the substance to be plated the negative electrode in a solution of a salt of the metal with which it is desired to plate it.

128. Electro-plating. The current used for electro-plating is commonly produced by a dynamo giving small pressure (three or four volts) and large current.

In silver-plating, for instance, a trough is provided containing a solution of potassio-cyanide of silver. The spoons, forks, or teapots which have to be plated are hung in the solution, and a wire from the negative pole of the dynamo is attached to them. The positive pole is attached to another piece of metal, which is also immersed in the trough. On working the dynamo the potassio-cyanide of silver is decomposed, and the silver is deposited on the articles to be plated.

QUESTIONS ON CHAPTER XIX.

1. What is Electrolysis?
2. What are its principal laws?
3. What is Clausius' theory of electrolysis?
4. What is the law of electro-chemical action?
5. What is the voltameter?
6. Describe the process of silver-plating.

CHAPTER XX.

ELECTRO-MAGNETS — DIAMAGNETISM AND MAGNE-CRYSTALLIC ACTION.

129. We have already stated that if an insulated wire be wound round a bar of iron or steel, and an electric current be sent round the wire, the core becomes a magnet with its marked end in the position in which the marked end of a permanent magnet would come to rest, if it were placed, free to turn, inside the helix instead of the iron core. When the core is of soft iron it becomes magnetized on the passage of the current, and loses its magnetism when the current ceases. When the bar is of steel, it resists the assumption of the magnetic state, but, when magnetized, retains its magnetism for an indefinite time after the cessation of the current.

The property of soft iron is taken advantage of in the construction of "electro-magnets," by means of which very great magnetic forces are obtained, which forces are under perfect control, for within certain limits the strength of the temporary iron-magnet is proportional to the strength of the current in the wire.

130. **Electro-magnets** are frequently made in the horseshoe form, and when in this shape usually consist of two parallel pillars of soft iron, connected at one end by a massive cross-bar of the same metal. The wires are wound on hollow brass reels, which can be lifted on and off the iron pillars.

The magnet shown in fig. 78 consists of a "horse-shoe," whose pillars are each 13 inches long and $2\frac{1}{4}$ inches diameter; the helices, which are 12 inches long and 5 inches external diameter, each contain about 1000 turns of insulated copper wire of about No. 18 gauge. The helices weigh about 35 lbs.

each. Such a magnet, if fixed with its poles downward, and

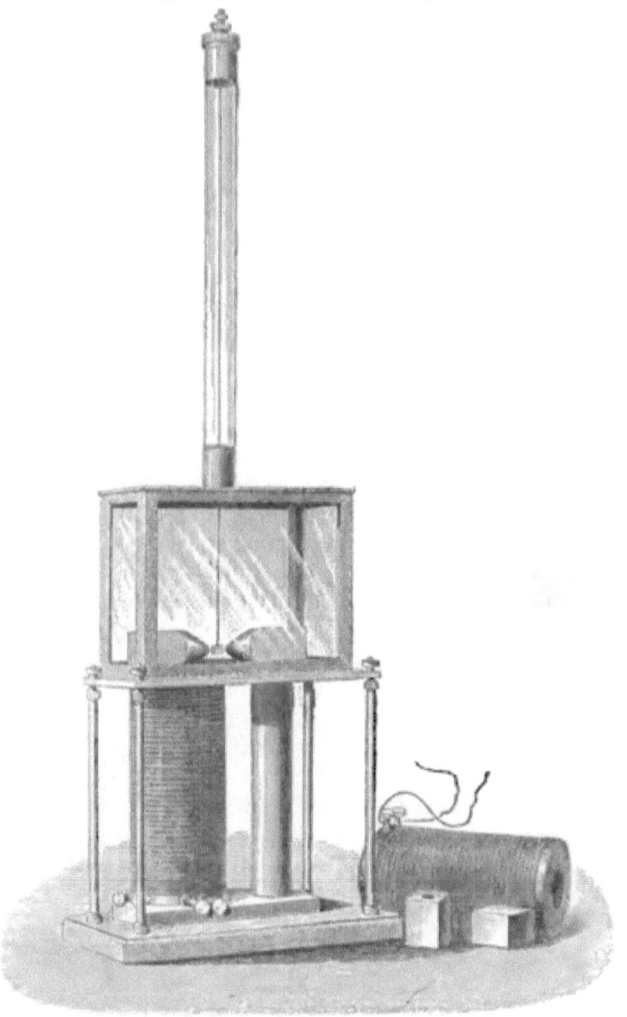

Fig. 78.

excited by a powerful current, would probably carry a weight of from one to three tons attached to the armature. Magnets of this size, however, are used for a different purpose. *Their magnetic action is so intense that it affects almost all known substances, in addition to those few commonly known as magnetic.* Some substances are attracted like iron, but others are repelled. The action on most substances is too feeble for the attraction or repulsion to be observed directly.

To observe the action, say, on a piece of glass, the magnet is placed with its pillars vertical and the cross-bar at the bottom. The free poles then project up about an inch above the helices. Two blocks of soft iron, which are called the movable poles, are placed upon the tops of the pillars. A body placed between these blocks is practically between two horizontal poles.

By sliding the blocks, the horizontal poles can be either approached close to each other or separated to a distance of some five inches. One end of each block is flat, so as to give the pole a vertical plane face about two inches square; the other ends are tapered to a blunt point. Either the flat or pointed ends can be turned towards each other. A body placed between the two pointed poles is subjected to more intense forces, while one placed between the flat poles is in a more uniform field of force; the action of the points being to concentrate the magnetic action in one place.

131. Torsion balance. In order to measure the magnetic actions of those bodies which are very feebly affected by the magnet, an apparatus called a torsion balance is used. It consists of a long fine thread, by which the body to be measured is hung. The top end of the thread is fixed to a movable circle carried on the top of the glass tube shown in fig. 78. On the magnetic forces holding a body in a particular position, it can be twisted out of that position by turning the circle and in twisting the thread. The amount of twist or *torsion* which has to be applied is measured by the number of degrees through which the top circle has had to be turned, and is a measure of the magnetic force which has been neutralized by the torsion. A homogeneous body being suspended between the poles by the torsion fibre will, if it is attracted, take up its position with its longest diameter pointing from pole to pole; and if it is repelled, its longest diameter pointing across the line joining the poles.

Definition.—The line joining the poles is called the axial line; the line at right angles to it, the equatorial.

132. Diamagnetism. *Definition.—Iron and similar bodies which are attracted by the magnet are called Ferromagnetic, or sometimes Paramagnetic bodies. Substances which are repelled are called Diamagnetic.*

The type of diamagnetic bodies is bismuth, which is repelled from a powerful magnet with considerable force. A small sphere ¼ inch in diameter **hung** by a thread, say, two feet long, between the pointed poles of a powerful magnet, may be repelled so as **to move** it as much **as a** quarter **of an** inch out of the vertical.

The phenomena of diamagnetism were first observed by Faraday,* **on a** piece of **a** particular kind of glass called heavy glass.†

In these **first** experiments **a** rod of heavy glass **was** suspended between the poles of **the** great horse-shoe magnet of the Royal Institution. It was found that the bar always **placed** itself equatorially, that **is, at right** angles **to** the line joining the poles, and that **it was in stable** equilibrium in **that** position. There **was also** another position of rest when the length of the bar **was exactly axial; but in** this case the equilibrium was **unstable, and on the least** displacement the bar moved to **the equatorial position.**

No **difference could be detected** between the ends of **the** bar; **the direction in which** either end pointed when **in stable** equilibrium **depended** solely on the direction **in which it was** displaced from the position of unstable equilibrium, **thus** showing **that** no permanent polarity, analogous to **the polarity of a** steel magnet, is acquired by the glass.

---

* " Isolated observations by Brugmans, Becquerel, **Le** Baillif, Saigy, and Leebeck had indicated the existence of a repulsive force exercised by the magnet on two or three substances, but these observations, which were unknown to Faraday, had been permitted **to remain** without extension or examination." Tyndall, " Faraday as a Discoverer," p. 110.

† **See** " Phil. Trans." 1846, and **in** the " Experimental Researches " **(2243).**

Faraday continued his experiments on a great number of substances, among which phosphorus showed the effect "as powerfully as heavy glass, if not more so." He also experimented on a great number of liquids.

Generally speaking, the distinction between para- and dia- magnetic substances is this:—Paramagnetics tend to move from weak to strong places of force, while diamagnetics tend to go from strong to weak places. Faraday found that almost all compounds of paramagnetic metals were themselves paramagnetic. Blood and yellow ferro-cyanide of potassium are, however, exceptions, as they both contain iron and are diamagnetic.

From numerous experiments Faraday deduced the following list, at one end of which is iron, the strongest paramagnetic; at the other, bismuth, the strongest diamagnetic. Metals nearest the neutral point 0 have the least action either way:—

Magnetic.	Diamagnetic.
Iron.	Bismuth.
Nickel.	Antimony.
Cobalt.	Zinc.
Manganese.	Cadmium.
Chromium.	Sodium.
Cerium.	Mercury.
Titanium.	Lead.
Palladium.	Silver.
Platinum.	Copper.
Osmium.	Gold.
	Arsenic.
	Uranium.
	Rhodium.
	Iridium.
	Tungsten.

0

It must be remembered that the diamagnetism of bismuth is very much smaller than the magnetism of iron.

The strength of the pole of an electro-magnet having an iron core may be as much as from thirty-two to forty-five times the magnetizing force, while with a bismuth core it would only be about—

$$\frac{1}{400,000} \text{ of it.}$$

133. Diamagnetic polarity. As soon as the facts of diamagnetism were established, the question arose—Are the effects observed due to simple repulsion, or is there a true diamagnetic polarity induced?—that is, Do diamagnetic bodies, when under the influence of magnetic forces, become temporary magnets, in the same manner as pieces of soft iron under the same circumstances, only with their poles in the opposite directions?

While the marked pole of the magnet induces, in a piece of soft iron in its neighbourhood, an unmarked pole at the side nearest to it and a marked pole at the farther side, we should, on the hypothesis of diamagnetic polarity, have, if a piece of heavy glass were substituted for the soft iron, a marked pole at the side of the heavy glass nearest to the marked pole of the inducing magnet, and an unmarked pole at the further side. Faraday and others made many unsuccessful attempts to determine whether or not diamagnetic polarity exists, and the matter remained uncertain until 1855, when Professor Tyndall* published an account of a series of experiments, by means of which he obtained a satisfactory proof of the polarity of bismuth.

The bismuth bar, $l\,l'$ (fig. 79), was suspended inside a

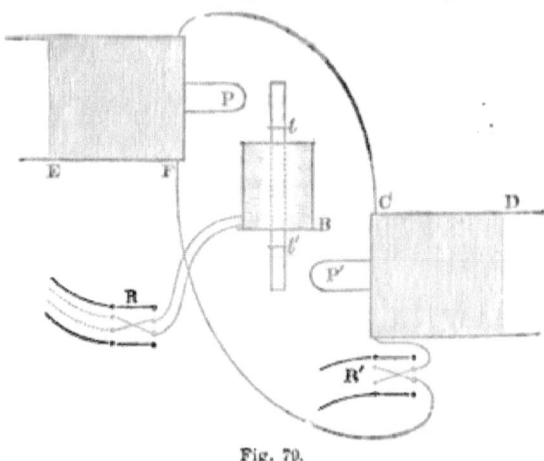

Fig. 79.

‡ Tyndall, "Diamagnetism and Magne-Crystallic Action," p. 130 (Longmans, Green, and Co., 1870); and "Phil. Trans." 1855, p. 33.

fixed helix, B, the tube of which was considerably larger than the bar; so that, within certain limits, the bar could swing like a galvanometer needle. Two single electro-magnets P P' were placed in the position shown. We know that if a bar of iron were substituted for the bismuth bar, it would be magnetized by the current in the helix, A B, and deflected to the right or left by the poles, P P', according to the directions of the currents in the helix and the magnet. When the current in A B was such that the polarity of the ends of the iron bar $l\, l'$, was the same as the polarity of the poles P P' opposite to them respectively, the bar was repelled; when the current in the helix was reversed, the bar was attracted.

On substituting a bar of bismuth for the bar of iron, exactly analogous effects were produced; but the directions of the currents which had produced attraction with the iron bar now produced repulsion with the bismuth and *vice versâ*, showing that the current in the helix, which produced a particular polarity in the iron bar, produced a *reversed polarity* in the bismuth. The commutators, R R', allowed every possible combination of currents to be tried.

In 1852 Professor Weber published a memoir, in which he discussed some of the mathematical consequences of diamagnetic polarity. He pointed out that—

"*The magnetism of two iron particles lying in the line of magnetization is increased by their mutual action, but, on the contrary, the diamagnetism of two bismuth particles lying in this direction is diminished by their mutual action.*

"The reverse occurs in both cases if the particles lie in a line perpendicular to the line of magnetization.

"From this it follows that to impart by a given magnetizing force the strongest possible magnetism to a given mass of iron, we must convert it into a bar as long and thin as possible, and set its length parallel to the line of magnetizing force; and that to impart the maximum diamagnetism to a given mass of bismuth, we must convert it into the thinnest plate possible, and set its thickness parallel to the line of magnetizing force."

Professor Tyndall afterwards made some much more delicate experiments, by means of which he established the diamagnetic polarity, not only of bismuth, but of many other substances both solid and liquid.

134. Magne-crystallic action. In all the experiments which we have hitherto described, we have considered the bismuth and other substances to be in a homogeneous state. When, however, they are in a heterogeneous or crystalline state, it was observed by Faraday that considerable differences are observed in their deportment under the action of powerful magnets.

The general law which determines the behaviour, in the magnetic field, of bodies whose density is not the same in all directions is, that *the magnetic axis induced in the body coincides with the line of greatest density.*

In crystals this line is parallel to the cleavage planes; that is, a diamagnetic crystal will set with its cleavage planes equatorial when suspended between the poles of a magnet, even when the diameter at right angles to the cleavage planes is considerably longer than that measured along them.

If, then, a bar of crystal, not too long, be cut so that its cleavage planes are perpendicular to the length of the bar, its behaviour, when suspended between the poles of a magnet, would be opposite to that of a homogeneous bar of the same shape, composed of a substance having the same magnetic properties.

The reason of this is that the magnetic induction parallel to the cleavage planes is so much stronger than that perpendicular to them, that the force tending to set the cleavage planes equatorial (if the crystal be diamagnetic) is stronger than that tending to set the length of the bar equatorial, in spite of the longer leverage which it has in that direction.

The first observations "on the crystalline polarity of bismuth and other bodies" were made by Faraday. His paper on the subject formed the Bakerian Lecture for 1849, and will be found in the "Phil. Trans." for that year.*

The subject was continued by Professor Tyndall, and his

* And "Exp. Res.," 2454, vol. iii. p. 83.

various papers on it are collected in his "*Diamagnetism and Magne-Crystallic Action.*"

He made some very interesting experiments on the effects of compression—that is, on the effects of producing artificially a "line of greatest density" in a particular direction. Perhaps the best of them was made accidentally. He was experimenting with the great electro-magnet of the University of Berlin, the copper helices of which alone weigh 243 lbs. A cube of bismuth was suspended between the poles, and the poles were accidentally brought rather too near together; their mutual attraction overcame the friction between them and the iron pillars on which they lay. They rushed together and crushed the bismuth between them, compressing it to about three-fourths of its former thickness. The poles having been separated and the bismuth extracted, it was boiled in hydrochloric acid to remove any trace of iron it might have acquired from the poles, and again suspended between them. The line of greatest compression at once set equatorial.

The poles were now purposely allowed to rush together, again pressing the bismuth along a line at right angles to the former line of compression. On being again cleaned and suspended, the new line of compression set equatorial. It was found, by repeating the experiment, that the direction of the magnetic axis could be changed as often as desired.

This experiment was the more remarkable as the bismuth had previously a natural crystalline structure, but the difference of density in the two directions, produced by the compression, was so much greater than that due to the direction of the cleavage planes, that the set was always determined by the direction of the artificial compression.

Professor Tyndall fills several memoirs with experiments to confirm and illustrate the law above described.

A paste made of wax and powdered bismuth is an excellent material from which to make artificial crystals by compression. They can also be made by compressing bread, if great care be taken as to the cleanliness both of the fingers and tools employed.

In these experiments it is usually necessary for the **experimenter** to wash his hands about every five minutes. The hands should be washed under a tap, so as to have a **constant** change of water, and dried with a "glass-cloth," which is not so liable to get dusty as an ordinary towel.

The following experiment of Professor Tyndall's on the construction of a model to show the effect of cleavage planes is of interest. Emery-paper is very strongly paramagnetic. Let **two** bars, each one inch long and half an inch square, be constructed of it. One, which we will **call** No. 1, is made by gumming together a sufficient number of strips, each one inch by half an inch, to make up the half-inch thickness; the other, which we will call No. 2, by gumming together a sufficient number of pieces, each half an inch square, **to** make up the inch length. On being suspended between the poles of a magnet, No. 1, which represents **a** crystal with its cleavage planes parallel **to** its length, sets axial. No. 2, however, in which the cleavage planes **are** perpendicular **to** the length, **sets** equatorial; that **is,** with its cleavage planes, and not its length, axial. **It** is **very** striking to see the behaviour of No. 2 when the magnet **is** powerful. The attraction **of** the mass **to the** nearest pole is **so** powerful that once, when the author **was** repeating the experiment, it broke a stout thread of sewing silk by which the bar was suspended, and **yet the** length **is** held very strongly in the equatorial line, **the** action being exactly that of **a** homogeneous bar **of** a strongly diamagnetic substance.

135. Effect of the surrounding medium. It is **found** that the medium in which the substance experimented **on hangs** between **the** poles affects the result of the experiment. For instance, a homogeneous bar of a feebly para**magnetic substance** will point axially in air or **a** vacuum, **but will** point equatorially if it **is** immersed **in a** strong solution **of** proto-sulphate **of iron.**

A series of experiments made by Faraday, and afterwards continued by Tyndall, have given the obvious and simple explanation of the matter. In order that the suspended body **may** take **up** any position, it has to displace an equal quantity of the surrounding medium from that position; but the magnetic force acts both upon the substance and upon the medium. If the action upon the substance is stronger than that upon the same quantity of the medium, the substance will take the same position as if it was in a vacuum. If, on the other hand, the magnetic action on the medium is greater than that on the substance, the medium will take the position to which the magnetic force tends to move it, and the substance will be displaced and will take the contrary position.

When this fact was established, it was suggested by Faraday that it might be possible to account for all the phenomena of diamagnetism without assuming the existence of a true repulsion, by supposing all space to be filled with a medium whose magnetic capacity was less than that of iron, but greater than that of bismuth, and that the supposed diamagnetic properties of bismuth might be accounted for by considering it merely as a feebler paramagnetic than the medium.

Professor Tyndall,* in a letter to Faraday, has pointed out that this explanation is not sufficient to account for the observed facts, and in particular that conclusions deduced from it as to magne-crystallic action are directly at variance with the results of experiment. The arguments in favour of the existence of true diamagnetic repulsion are unaffected by the consideration of the effects of the media-surrounded bodies under the action of magnetic forces.

Questions on Chapter XX.

1. Describe and sketch a large electro-magnet with torsion apparatus.

2. What is the difference between diamagnetic and paramagnetic bodies?

* "Diamagnetism," p. 213.

3. What are approximately the relative strengths of the paramagnetic action of iron and the diamagnetic action of bismuth?

4. If a ball of bismuth be hung nearly, but not quite, between the pointed poles of a powerful magnet, will it be drawn in between them or repelled further from them? How would a ball of iron be acted on in the same circumstances?

5. What is meant by diamagnetic polarity? Describe an experiment for proving its existence.

6. State the law of magne-crystallic action.

7. Show how artificial crystals can be made.

8. A thin glass tube filled with a weak solution of sulphate of iron is suspended horizontally between the poles of a magnet in a glass trough. It points axially, and behaves generally as a paramagnetic body. What happens on the trough being filled up with a much stronger solution of sulphate of iron? Explain the result.

CHAPTER XXI.

THE INDUCTION COIL.

If a magnet be placed inside a coil of wire and suddenly withdrawn, a momentary current of electricity will be produced in the coil, and its electro-motive force will be greater the more suddenly the magnet is drawn out.

If, instead of removing the magnet, we destroy its magnetism, we find, as we might expect, that a current is still induced in the coil. If, instead of a steel magnet, we use an electro-magnet, we can, by starting and stopping the current, make and destroy the magnetism much more suddenly than we can insert and withdraw a steel bar; and so, on the destruction of the magnetism, we shall induce a current in a surrounding coil having a much greater electro-motive force than could be produced by the withdrawal of a steel magnet of equal strength.

136. The induction coil is an instrument in which advantage is taken of the fact of electro-magnetic induction, to convert the electricity of the voltaic battery current, which has large chemical, heating, and magnetic effects, but of which the greatest difference of pressure between its different points is comparatively very small, into electricity with much less chemical and magnetic power, but of which the difference of potential at different points in the circuit is enormous.

In order to obtain some idea of the difference between the electric pressures given by a battery through the medium of a coil and those of the largest batteries directly, we may note that Messrs. De La Rue, Müller, and Spottiswoode,[*]

[*] "Proc. Roy. Soc." vol. xxiii. 1875, p. 357.

found that with 1080 chloride of silver cells, it was only possible to obtain a spark, whose length varied from $\frac{1}{253}$ inch to $\frac{1}{150}$ inch, while even small induction coils give sparks of over an inch with one or two cells—and Mr. Spottiswoode's great coil, the largest ever constructed, gives, with fifty Bunsen cells, a spark of $42\frac{1}{7}$ inches.

The induction coil consists mainly of an electro-magnet placed inside a coil of fine wire and an apparatus for magnetizing, and demagnetizing the electro-magnet as rapidly as may be desired.

The following is the general outline of its construction. A core of soft iron is covered by an insulating material, and round it are wound a few layers of stout insulated wire, in the form of a helix. This helix is called the *primary coil*. Outside it and well insulated from it by means of a thick ebonite tube, is wound a very great length of very fine insulated wire, forming a great number of layers, each of which consists of a great many windings. The fine wire is called the *secondary coil*.

The *core* is usually made of a bundle of iron wires, as they will demagnetize more quickly than a solid bar.

The ends of the primary wire are connected to a battery, the ends of the secondary to discharging points separated by a greater or less interval of air. When the current is made or broken in the primary circuit, a certain difference of potential is caused by induction along each portion of the secondary circuit, each winding being acted on by the iron core, and by the portions of the primary circuit near to it.

There are a great number of windings of the secondary, and a separate difference of potential is produced in each. These being all added together produce a very great difference of potential at the discharging points, which difference is sufficient to break down the resistance of the air between the points.

Fig. 80 represents a coil, by Apps, in the possession of the author, which gives a spark of seventeen inches in air, and has twenty-two miles of secondary wire.

It is very necessary that contact should be broken as sud-

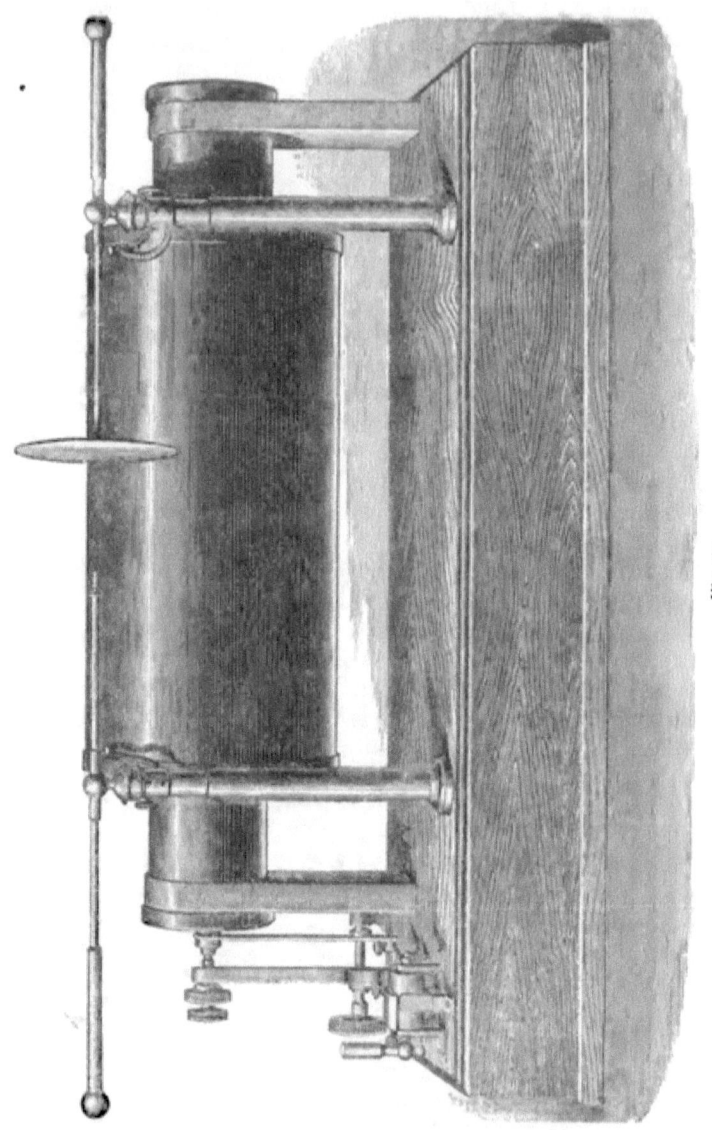

Fig. 83.

denly as possible, in order that the differences of potential produced at different points of the secondary circuit may all act at once in producing a great difference of potential at the extremities.

It is found that, except in very large coils, sparks are only produced on breaking the circuit, and even in very large coils the spark produced by closing is much feebler than that produced by opening. This may be due to the fact that the magnetization of the iron core takes longer to rise to its maximum value than to sink from its maximum to zero.

137. **The condenser.** This is a very important portion of an induction coil. It consists of a number of sheets of tinfoil, separated by mica, gutta-percha, or paraffined paper. The 1st, 3rd, 5th, 7th, &c., sheets are connected to one end of the primary wire, the 2nd, 4th, 6th, 8th, &c., to the other end. When the circuit is broken, the extra current, induced in the primary wire by breaking, is in the same direction as the primary current, and therefore tends to prolong the magnetization of the core. When a condenser is used, the extra current spends itself in charging it. The condenser then, instantly discharging itself, sends a current in the reverse direction round the core, and at once demagnetizes it. The condenser is usually placed in the base of the coil.

138. **Contact breakers.—The vibrator.** Various methods are used to make and break the circuit.

The form of contact breaker which is universally used for small coils is called the vibrator (fig. 81).

It consists of a piece of iron, which is supported near one end of the core by a brass or steel spring, which tends to pull it away from the core, and to force a piece of platinum soldered on to the back of the spring against a platinum pointed stop. The position of the latter can be regulated by a screw. The primary current passes from the stop to the spring.

A second screw regulates the tightness of the spring. This tightening screw works in an ivory collar for insulation, and usually has an ebonite head.

When the current passes, the iron core becomes a magnet, and, pulling the spring forward, separates the platinum

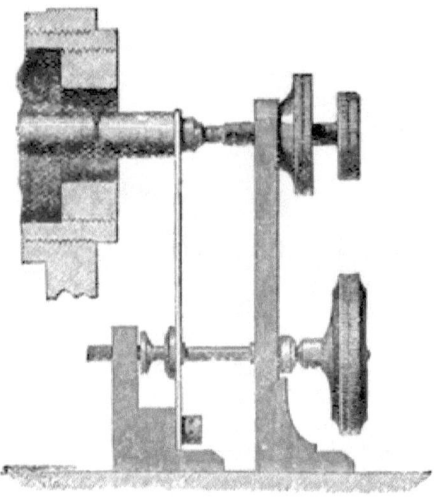

Fig. 81.

points and breaks the current. This, at the same time, destroys the magnetism of the core, and the spring flies back and completes the circuit, when the process is repeated, and thus a constant vibration is kept up, the action being like that of an electric bell. When the spring is weak, the current is broken at a time when the core has but small magnetic strength, and a feeble induction current is produced. By tightening the spring we may arrange the apparatus so that the current is not broken till the core has received nearly its maximum of magnetization, and so a much stronger current of induction is generated.

Various other breaks for the primary circuit are in use, but need not be described here.

139. Mr. Spottiswoode's coil. A description of this, the largest coil ever constructed, will be found in the *Philosophical Magazine* for January, 1877.

It was made by Mr. Apps. The total weight of the coil

is 15 cwt.; its length 4 feet; its external diameter 20 inches. It is supported on two massive pillars at its ends, while a central pillar, adjusted by a screw, provides against any bending that may occur.

There are two primaries, one of which may be replaced by the other by two men in the course of a few minutes. The one to be used for long sparks, and for most experiments, has a core consisting of a bundle of iron wires, 44 inches long, by 3·56 inches in diameter, and weighing 67 lbs.

The copper wire of this primary is 660 yards long, ·096 inch (nearly $\frac{1}{10}$ inch) in diameter, and has a total resistance of 2·3 B.A.U., with a conductivity of 93 per cent. It contains 1344 turns, wound in six layers; its weight is 55 lbs., and total length 42 inches. The other primary, which is intended for a short thick sparks, has a thicker core, weighing 92 lbs., and the wire is wound in double strands, so as to give much less resistance.

The secondary coil consists of no less than 280 miles of wire forming a cylinder 37·5 inches long, and 20 inches external diameter. The total resistance is 110200 B.A.U. It is wound in four sections. The diameter of the wire used for the two central sections being ·0095 (nearly $\frac{1}{100}$ inch) and those of the external a little thicker.

Using the smaller primary, this coil gave :—
 With 5 quart cells of Grove a spark of 28 inches.
 With 10 ,, ,, ,, 35 inches.
 With 30 ,, ,, ,, 42½ inches.

A spark of 42½ inches is by far the largest that has been obtained by any electrical apparatus whatever.

Question on Chapter XXI.

Describe the induction coil and give the theory of its action.

CHAPTER XXII.

ON THE DISCHARGE OF THE INDUCTION COIL AND DISCHARGE GENERALLY.

140. Discharge. The phenomena about to be described have, except where the contrary is stated, been observed with a coil capable of giving 17 inches spark. When the discharge is taken between a point and a disc separated to the maximum distance over which a spark will pass, and the point is made positive, the discharge consists of a zigzag line of bluish white light, and is accompanied by two distinct sounds; one, a crackling which, when a slow working break is used, resolves itself into a separate sharp report like that of a small percussion cap at the instant of each discharge; and the other a hissing sound, caused by the appreciable time required by the twenty-two miles of secondary wire to discharge itself.

This last effect is more particularly noticeable when the points are near. In Mr. Spottiswoode's great coil, when the points are placed four or five inches apart, the hissing discharge lasts some two seconds or more. When the point is positive, the discharge always takes place near the centre of the disc, but seldom twice in exactly the same spot. When the current is reversed so as to make the disc positive, then with the same battery power only a much shorter spark can be obtained, and it takes place between the point and the edges of the disc. I am not aware that the reason of this difference is known.

When the battery power is weakened, or the vibrator spring relaxed so that the coil will not give quite its full spark, then, if the discharging rods be separated to rather a greater distance than that over which the spark can **pass**, the space between will, **if** the room be darkened, be observed to glow with a faint **blue** light, extending for some two or three inches in all directions round the line joining the poles. This has been called the brush discharge.

When the points **are** put at about the maximum distance over which a spark can pass, the spark and brush discharge are frequently observed together.

When the points are brought within some two or three inches of each other, **the** discharge **is** thicker but nearly silent, and is surrounded by a mass of yellow flame of some ½ inch **to** ¾ inch **thick**. This **is** caused chiefly by the combustion of the sodium in the air. It can be blown away by a **current** of **air**, leaving the spark unaffected. A candle **or** piece of paper can **be** lighted **at** it.

The discharge from even **a** very small coil, if taken through the body, produces violent pain and muscular contraction, **the** patient being usually unable to leave go of **the** electrodes. The discharge of such a coil as the 17-inch **would** probably be fatal; indeed, that of a coil giving **a** ¼**-inch** spark is the maximum that could be taken with safety.

If an ordinary looking-glass, **about 15** inches by 10 inches, be set with **its** back against **the disc of a** 17-inch coil, and the **point be put near the middle of** the glass **face, a very beautiful effect is observed. The** discharge appears to **strike the** glass, and breaks in a kind of spray of fire, streaming **in** every direction to the edges, whence it is conducted by the mercury **and** wooden back, to the disc.

By placing the points of the secondary at opposite sides **of pieces of** plate-glass, and surrounding the terminals with **cement,** considerable thickness of plate-glass **can** be perforated. With **the** great Polytechnic coil, five **inches of** glass **were** perforated.

141. Secondary condenser. If wires **be led from** the secondary terminals **to the opposite** coatings of **a** Leyden jar, **or** other condenser, **the character** of the discharge **is** changed. The discharges are **not** quite **so** frequent, as the **jar** has **to be** charged between each spark, but the spark is of a dazzling white, and instead of the usual smart crackle of the impinging sparks, a series of deafening reports are heard. If a slow working break is used, so that there is an interval between each discharge, the metal disc is heard to ring after each spark, as if it had been struck by a hammer.

The difference between an ordinary discharge and a discharge with a secondary condenser, may be illustrated by the following water experiment. Let water run out through a tap in a series of drops, following very closely after each other; each makes a little splash as it falls; this may represent the ordinary discharge.

Now place under the tap a little bucket, so pivotted that it will tip over when full, and then right itself. The stream will be interrupted while the bucket is filling, and then there will be a large splash when it turns over; and instead of a large number of small splashes occurring in each minute, there will be a small number of large ones. This represents the discharge with the condenser.

We note that the same quantity of water or of electricity is used in the respective experiments with or without the bucket or condenser.

When a spark is taken between a point and a polished plated disc, each discharge causes a minute dot on the bright surface, which cannot be rubbed off. It is due to the volatilization and burning of a small portion of the silver.

The noise caused by a $\frac{1}{2}$-gallon jar, with a 17-inch coil, is sufficient to make it necessary to shout in order to be heard.

At the same time, as the strength of the spark is increased, the length is decreased. With a large jar in my possession, which contains eleven gallons, and has $7\frac{1}{2}$ square feet of coated surface, the maximum spark which the 17-inch coil will give is something under 1 inch. With small jars, the length of the spark is limited by the size of the uncoated portions of the jars, as when the points are separated by more than a certain distance the spark springs round the glass. With a $\frac{1}{2}$-gallon jar of the shape usually sold, about 5 inches spark can be obtained.

The jar discharge will perforate paper, but **not** ignite it.

142. Induction coil and magneto-electric or "dynamo" machine. In November, 1879, Mr. Spottiswoode* published an account of some experiments, in which a 20-inch coil was excited by means of the current produced by De Mériten's magneto machine, worked by a 3½ horse-power gas-engine. This machine gives *alternate* currents whose direction is reversed about $16 \times 1300 = 20,800$ times per minute, and therefore no contact breaker or primary condenser are required.

This method of working gives secondary currents having great "quantity."

With a 20-inch coil the spark is about 7 inches long, and has the "full thickness of an ordinary cedar pencil." The discharge is extremely regular, and can be used for spectroscopic purposes without a secondary condenser.

143. Discharge in rarefied air. When the discharge either of a coil or electrical machine is passed through a tube, or other vessel connected to an air-pump, it is found that as the pressure diminishes, the length of spark which can be obtained increases. A great many experiments have been made to determine the exact law according to which the spark-length increases as the pressure diminishes.

In the experiments made by M. Masson, the spark passed either between two balls in the air, or between two similar balls inside a globe in which a more or less complete vacuum could be produced. The distances between the balls could be varied, as well as the pressures. Within the limits of his experiments, M. Masson found that the length of spark was inversely proportional to the pressure. The greatest length of spark which he used was 11·1 millims.

144. Gordon's experiments. At the Dublin Meeting of the British Association the present writer gave an account of some recent experiments which he has made on

---

* With the great coil made by Mr. Apps for the Polytechnic Institution, which gave 29 inches spark, a shock was administered to a rabbit without causing death. It is probable, however, as the rabbit was much less than 29 inches long, that the greater part of the discharge passed round him through the air and not through his body.

the subject. In them an attempt has been made to determine the ratio of the spark-length to the pressure for distances ranging from 6 inches to 30 inches by means of one and the same apparatus. The experiments differ from any former experiments with which the author is acquainted, in the fact that an induction coil was used as the source of electricity instead of an electric machine.

The coil was the 17-inch coil already mentioned.

It was worked by 10 quart cells of Grove's battery arranged in series. It was provided with a vibrator and with a clock contact-breaker, either of which could be used.

The Air-Pump was of the ordinary Tait's construction.

The Discharging Tubes.—These consisted of two cylindrical glass tubes about 4 feet long and nearly 3 inches diameter. At one end of each was a tap, the brass pipe from which ended in a ball which formed one of the discharging terminals. Holes in the side of the brass pipe admitted the air from the tap to the tube. At the other end of each tube was a stuffing-box, in which a brass rod slid; at the end of the brass rod was a point which could either be placed in contact with the ball or withdrawn some 3 feet from it. The end of the rod was kept always in the axis of the tube by means of three little glass arms, which were inserted into an ebonite collar fixed on the discharging rod a little behind the point. The two tubes were supported in a horizontal position, parallel to each other and about 18 inches apart, on four ebonite legs about 18 inches high. The tubes were joined to the air-pump by means of the pipes and taps shown in fig. 82, which were so arranged that the tubes could be quickly connected to each other, to the external air, to a gas-holder, or to the pump. Between the tubes and the pump the metal pipe was cut, and a piece of glass tubing about 18 inches long, well varnished with shellac, was inserted, so that the electricity might not pass to earth through the pump.

When the tubes were shut off from the pump, air could always be let into the glass pipe to prevent the discharge passing to earth inside it, as it would do at low pressures. The distance between the point and

Electric Discharge. 173

ball in each tube was measured as follows:—They were placed in contact, and an ink-mark was made on the

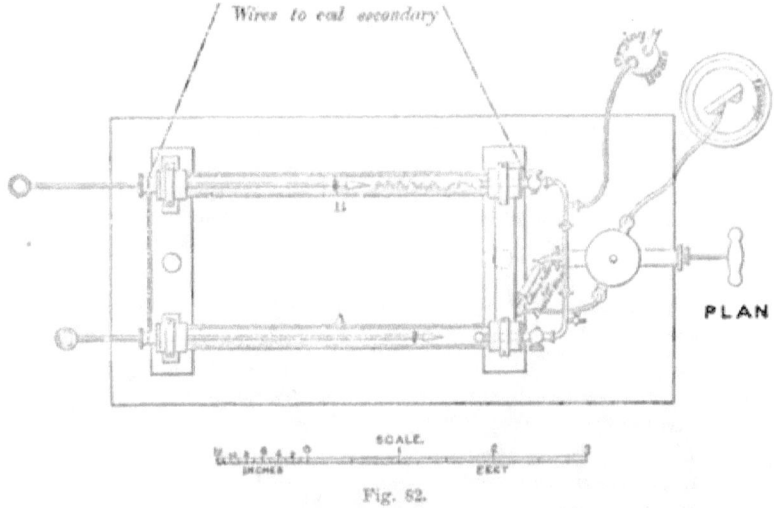

Fig. 82.

discharging rod just outside the collar of the stuffing-box. When the rod was slid out, the distance of this mark from the collar was equal to the distance between the point and ball. The pressure was given by a U-gauge, about 4 feet high, attached to the air-pump at one end, open to the air at the other.

The pressure P was given by the formula—

$$P = \{\text{height of barometer}\} - \{\text{difference of level of mercury in the two arms of the U}\}.$$

Before being admitted into the tubes, the air was dried by being drawn through sulphuric acid. When it was desired that the pressure of the air in the tube should equal that of the external atmosphere, air bubbled through the acid as long as the difference of pressure inside and outside the tube exceeded that of the inch of acid which had to be displaced, and then the tap was opened direct to the outside air. The external diameters of the tubes were about 2·94 and 2·76 inches respectively, and the diameters of the balls ·94 and ·92 inch.

In the experiments, one of the tubes (A) was left open to the atmosphere, and its discharging point placed at a standard distance either 6, 8, or 10 inches from the ball; and the other tube (B) being nearly exhausted, experiments were commenced at the low pressure, and then a little air was let in between each observation. The tubes were so connected to the coil that the discharge would pass in whichever tube offered least resistance. The discharging distance in B was then varied and adjusted to the *shortest* distance, which caused the whole discharge to pass in A. This distance having been noted, the points in B were brought nearer together till they reached the *longest* distance at which the whole discharge passed in B.* The mean of these two distances was taken as the distance which, at the pressure then being worked with, interposed in B a " resistance " equal to that of the standard length in A of air at the pressure of the atmosphere.

Let us call this mean, " mean B spark." Now, if the law that the spark-length is inversely proportional to the pressure holds, we should have for the same series of experiments—

$$\{\text{mean B spark}\} \{\text{pressure in B}\} = \text{const.};$$

and to compare different sets made with different distances in A and with the barometer at different heights, we should have—

$$\frac{\{\text{mean B spark}\} \{\text{pressure in B}\}}{\{\text{distance in A}\} \{\text{height of barometer}\}} = \text{const.}$$

If the two tubes and the discharging points were precisely alike, this constant would be unity. Any slight difference in the shape of the points and balls would cause it to differ from unity, but would not affect its constancy.

145. Results. (1) From an air pressure of about 11

---

* The fact that the discharge only divided itself between the two tubes, when the "resistances" were almost equal, confirms Mr. De La Rue's discovery (see my "Electricity," vol. ii. page 152) that disruptive discharges do not obey Ohm's law.

ball in **each tube** was measured as follows:—**They were** placed **in contact,** and **an** ink-mark was made **on the**

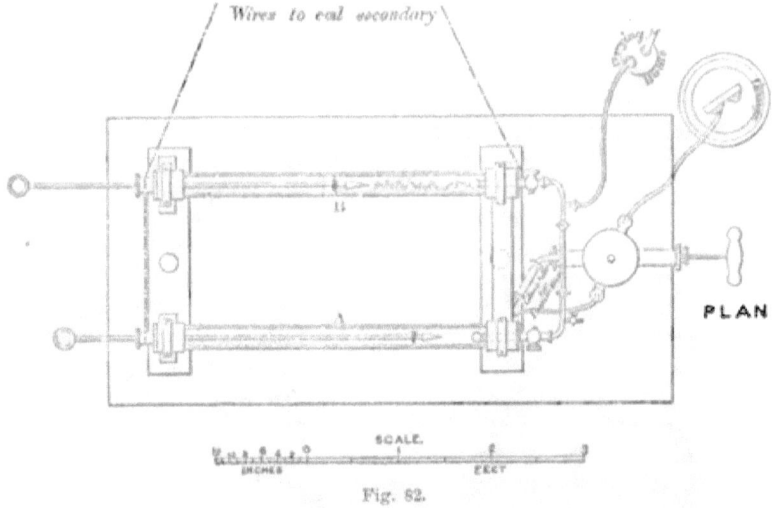

Fig. 82.

discharging rod just outside the collar of the stuffing-box. When the rod was slid out, the distance of this mark from the collar was equal to the distance between the point and ball. The pressure was given by a U-gauge, about 4 feet high, attached to the air-pump at one end, open to the air at the other.

The pressure P was given by the formula—

P = {height of barometer} − {difference of level of mercury in **the two** arms of the U}.

Before being admitted into the tubes, the air was dried by being drawn through sulphuric acid. When it was desired that the pressure of the air in the tube should equal that of the external atmosphere, air **bubbled** through the acid as long as the difference **of pressure** inside and outside the tube exceeded that of the **inch of** acid which had to be displaced, and then the tap **was opened** direct to the outside air. The external diameters **of the tubes** were about 2·94 and 2·76 inches respectively, and the diameters of the balls ·94 and ·92 inch.

In the experiments, one of the tubes (A) was left open to the atmosphere, and its discharging point placed at a standard distance either 6, 8, or 10 inches from the ball; and the other tube (B) being nearly exhausted, experiments were commenced at the low pressure, and then a little air was let in between each observation. The tubes were so connected to the coil that the discharge would pass in whichever tube offered least resistance. The discharging distance in B was then varied and adjusted to the *shortest* distance, which caused the whole discharge to pass in A. This distance having been noted, the points in B were brought nearer together till they reached the *longest* distance at which the whole discharge passed in B.* The mean of these two distances was taken as the distance which, at the pressure then being worked with, interposed in B a " resistance " equal to that of the standard length in A of air at the pressure of the atmosphere.

Let us call this mean, " mean B spark." Now, if the law that the spark-length is inversely proportional to the pressure holds, we should have for the same series of experiments—

$$\{\text{mean B spark}\} \{\text{pressure in B}\} = \text{const.};$$

and to compare different sets made with different distances in A and with the barometer at different heights, we should have—

$$\frac{\{\text{mean B spark}\} \{\text{pressure in B}\}}{\{\text{distance in A}\} \{\text{height of barometer}\}} = \text{const.}$$

If the two tubes and the discharging points were precisely alike, this constant would be unity. Any slight difference in the shape of the points and balls would cause it to differ from unity, but would not affect its constancy.

145. Results. (1) From an air pressure of about 11

---

* The fact that the discharge only divided itself between the two tubes, when the " resistances " were almost equal, confirms Mr. De La Rue's discovery (see my " Electricity," vol. ii. page 152) that disruptive discharges do not obey Ohm's law.

inches up to that of the atmosphere the spark-length for the same electric pressure is approximately inversely proportional to the air pressure.

(2) **No** law can be said to be **more** than approximately true; **for** when **the** density has almost reached the discharging limit, **any** slight accidental circumstance, such as the presence of a grain of dust, a little burning of the point by the **last** discharge, &c., will cause the discharge to take place. Professor Clerk Maxwell has compared this experiment **to the** splitting **of** a piece of wood **by a** wedge. It is possible to determine the average pressure on the wedge which will split the wood; but in any particular experiment it is impossible **to say** that **the** wood **will split exactly at** that pressure.

(3) When **the pressure is** diminished below 11 inches, we find that the **spark produced** by a given electro-motive force is **much shorter than** is required by the above law, or that the electric pressure required to produce a spark of given length **is at** low pressures greater than that required by the law.

Also that at a constant **air** pressure greater **electric** pressure per unit length of air is required **to** produce a spark at short distances than **at long ones**; or, if **we** adopt Faraday's view **that the tension exists in** every part of the air, they **show that air in a** thin **stratum has greater** strength than when **it is in a thick one.**

In fact, we may say that greater **electric pressure** per unit quantity **of air is** required to produce a **spark** with few **air particles between** the poles than with many.

We shall discuss some possible reasons for this **in a later portion** of this book.

146. Discharge in **different gases.** On May 17, 1877, Messrs. De La Rue **and Müller** stated that the length **of spark given by a battery at ordinary** atmospheric pressures **in the following gases is the longest** in the order in which **they are enumerated—hydrogen, nitrogen,** air, oxygen, **carbonic acid—it being nearly twice as long in** hydrogen as **in air.**

The spark does not appear to be dependent on the specific

gravity of the gas, but may have some relation to its viscosity.

147. Vacuum tubes. When the pressure of the air is less than about 15 inches of mercury, the appearance of the discharge changes considerably. The whole gas within the tube glows, and if the light be examined with a spectroscope, it will be seen to give the characteristic spectrum of the gas.

When the exhaustion is continued by a mercury pump till the pressure is only a very small fraction of a millimetre, the whole tube is filled with a bright light, of which the colour varies with the nature of the residual gas in the tube.

If any fluorescent substances are placed in the tube or surrounding it—if, for instance, a portion of the tube passes through a solution of sulphate of quinine, or part of the glass be coloured with uranium, they will glow with their characteristic colours when illuminated by the electric discharge.

In these "vacuum tubes," as they are called, the electrodes usually consist of platinum or aluminium wires, passed through the glass, which is then fused round them.

Platinum is particularly suitable for this purpose, because its expansion rate is about the same as that of glass, and, therefore, it does not crack out of the glass on cooling. A little opening being left at one side of the tube, the glass is drawn off into a capillary tube and attached to a Sprengel air-pump.

When the exhaustion has been carried as far as required, the capillary tube is heated in a blowpipe flame till it softens, when it is drawn off and so closed, a process which is assisted by the pressure of the external air.

148. Effect of magnets. It is found that the discharge in rarefied gas is attracted and repelled by a magnet in the same way as a wire carrying a current subject only to the differences caused by the fact that the wire is rigid and the discharge flexible.

149. Striæ. It is observed that when the vacuum tube is made somewhat narrow, as, for instance, when it is of the

Striæ.   177

form shown in fig. 83 that in the narrow part the stream

Fig. 83.

of light is not continuous, but is separated into a number of discs of light.

Under certain circumstances these discs are also observed in larger tubes. They are called "striæ" or stratifications. Their cause is not yet fully understood. A great deal has been done towards investigating them, and a summary of the present state of our knowledge of the subject will be found in my "Electricity." Figs. 84, 85, show some of them.

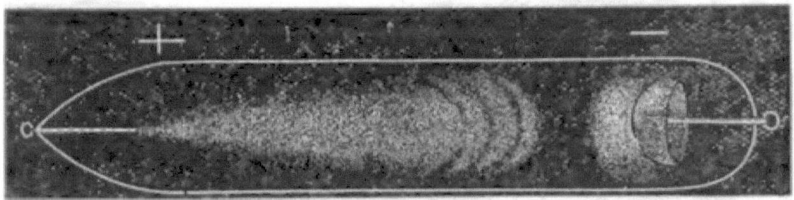

Fig. 84.

Fig. 85.

Nothing more beautiful of its kind can be imagined than the appearance of these masses of soft light hovering self-supported in the dark tube, sometimes at rest, and sometimes chasing each other from one end of the tube to the other, each as it comes to one terminal disappearing, while a new one appears at the other terminal.

N

Questions on Chapter XXII.

1. Describe the discharge of a large induction coil.
2. What is the effect of using a secondary condenser?
3. What difference is there in the discharge when it occurs in rarefied air?
4. What effect in the length of the discharge is produced as the pressure of the air is diminished?
5. What is a vacuum tube?
6. What are striæ?

CHAPTER XXIII.

ON THE NATURE OF MATTER.

150. We have now advanced sufficiently in our study of electrical science to be able to comprehend some electrical illustrations of the theory of matter, which we shall presently consider; and we need no apology for digressing from the study of electricity proper to the general theory of matter, for it is only by its bearing on the general theory of the constitution of the universe, and the laws by which the latter is ruled, that any particular branch of study, such as electricity, can be said to have a theoretical and scientific interest as distinguished from that due to its practical and commercial importance.

Matter is ordinarily known to us in three forms—solid, liquid, and gaseous.

According to the theories now generally accepted, all matter consists of "molecules," or very small particles packed in various stages of closeness. When the molecules are packed very tight, so that they cannot move past each other, or change their relative positions, the matter is said to be solid. When they are less closely packed, so that they can move past each other, but yet are closely linked by their mutual attraction, the matter is said to be liquid; and when they are still less closely packed, so that they do not greatly attract each other, but can move about independently, the matter is said to be gaseous. In the ordinary gaseous condition the molecules running about in their independent courses are constantly colliding with and rebounding from each other,

Matter in its three states may be represented by a crowd of people in various conditions. First, solid matter may be represented by a dense crowd so tightly packed, that, although each person can pant and breathe independently, or even turn partly round, and although the whole crowd can sway backwards and forwards, yet no one person can move from one part of the crowd to the other, or change his position with regard to other persons in it.

The liquid state may be represented by a much less dense crowd, where every one wishes to move about but is bound by a rule to always have at least one hand joined to a hand of some other person; in fact, where every one is dancing a sort of "grand chain." Any person (or molecule) can then circulate through the room, but is never independent of the pull of his neighbour's hand.

The gaseous state may be represented by considering the number of persons in the crowd to be still further reduced and to be blindfolded, and to keep their hands in their pockets and compelled to run violently about, colliding with each other and with the walls of the room. We see in this case that if the room is large and the number of persons great, so that the law of averages may hold, the number of persons who run against each square yard of wall per minute is constant, or that the pressure on the walls of the room is constant.

We notice that there is no hard-and-fast line of demarcation between these three states, but they fade gradually one into the other as the density of the crowd varies, and this is the case with the changes of matter from one state to another as the density of the molecules varies.

For instance, every one has seen a candle on a warm day bend over by its own weight. Although the wax never becomes liquid, its molecules obtain sufficient mobility from the rise of temperature to move slowly over each other under the influence of gravitation in the whole mass. The motion of glaciers is another instance of this mobility of viscid solids.*

151. The size of molecules can be approximately

* See Tyndall's "Forms of Water."

measured by experiments on the colours of the thin films of soap-bubbles, which films are of a thickness comparable with the length of the light waves, and cause new colours by assisting or retarding the motion of part of the light by less than one wave length.

We cannot in this book go into details of experiments, we must be content with stating the results arrived at. For details the student is referred to "The New Chemistry," * by Professor J. P. Cooke, of Harvard, of which we may here say that it combines the interest of a romance with the logical accuracy of a scientific treatise.

The results which have been obtained are that the size of a molecule of water is between $\frac{1}{250,000,000}$ and $\frac{1}{500,000,000}$ of an inch.

To get some idea of the degree of "coarse-grainedness," as we may call it, that these figures represent, we may add that if we consider a drop of water the size of a pea magnified to the size of the earth, each molecule being magnified to the same extent, then the magnified structure would be coarser grained than a heap of small shot, and less coarse grained than a heap of cricket-balls.

152. Radiant matter. In our illustration of the gaseous state we have considered the crowd although thin enough to allow each person to move about independently, yet so thick that each person in crossing the room will collide with a great many others before he reaches the other side, and as at each collision we suppose his direction of motion to be changed by the rebound, the chances are much against any particular person who starts in a given direction arriving at the other side of the room in that direction.

If, however, we still further reduce the number of persons in the room, we shall reach a state where, although there will still be collision, yet the majority of the persons will arrive at the other side of the room in the same direction as that in which they started, and if we suppose an intelligent man with his eyes not blinded to start a number of persons at once in the same direction, say all towards a certain door at the other side of the room, the majority will arrive there,

* International Scientific Series. Kegan Paul and Co.

and may perhaps burst open the door by their rush. No greater force is used than in the case of the denser (gaseous) crowd, but the smaller number of collisions enables the rush to be directed, and a greater bombarding force to be concentrated on a particular spot.

153. Crookes' experiments. Professor Crookes has realized this condition with matter. By means of improved air-pumps he has been able to so reduce the quantity of gas in a globe that the *free path*, as he calls it, of each molecule is great in comparison with the number of collisions it undergoes in crossing the vacuum, and by means of electricity he has been able to impel these molecules in definite directions, and produce definite "bombardment" effects.

To produce these effects he reduces the quantity of air or gas in his globe to about one-millionth of what it would be at ordinary pressures, the quantity in vacuum tube, which produces the effects of ordinary discharge described in the last chapter (**148**), being about $\frac{1}{500}$ of that at ordinary pressures. To matter in this highly attenuated state Mr. Crookes gives the name of *radiant matter*.

We will now proceed to describe some of Mr. Crookes' experiments.

In these experiments vacuum tubes are prepared similar to those we have already mentioned, but, as we have said, very highly exhausted. For gas in an ordinary vacuum tube the *mean free path* of each molecule, that is the average distance it travels without collision, is about $\frac{1}{500}$ inch; but in Mr. Crookes' tubes the mean free path is increased to about one inch.

When the current of an induction coil is sent through the tube, the molecules are shot off from the negative pole in straight lines, and the majority of them cross the tube in straight lines and strike whatever is opposite to the negative pole, like the bullets of a machine-gun. This bombardment produces effects which we shall now consider.

154. Radiant matter exerts powerful phosphorogenic action where it strikes. The first and one of the most noteworthy properties of radiant matter discharged from the negative pole is its power of exciting phosphorescence when it strikes against solid matter. The number of bodies which respond luminously to this molecular bombardment is very great, and the resulting colours are of every variety. Glass, for instance, is highly phosphorescent when exposed to a stream of radiant matter. Fig. 86 represents

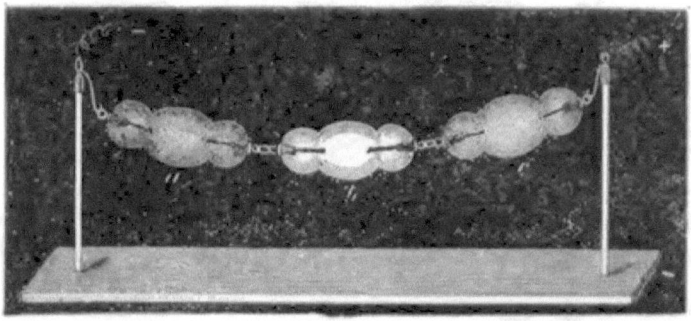

Fig. 86.

three bulbs composed of different glass : one is uranium glass (a), which phosphoresces of a dark green colour; another is English glass (b), which phosphoresces of a blue colour; and the third (c) is soft German glass—of which most of the apparatus used in the lecture was made—and phosphoresces of a bright apple-green.

Many other substances are also phosphorescent under the influence of radiant matter.

When luminous sulphide of calcium, prepared according to Mr. Ed. Becquerel's description, is exposed even to candle-light, it phosphoresces for hours with a bluish white colour. It is, however, much more strongly phosphorescent under the influence of the luminous discharge in a good vacuum.

The rare mineral Phenakite (aluminate of glucinum) phosphoresces blue; the mineral Spodumene (a silicate of aluminium and lithium) phosphoresces a rich golden yellow; the emerald gives out a crimson light. But Mr. Crookes finds that, without exception, the diamond is the most sensitive substance he has yet met for ready and brilliant phosphorescence. A very curious fluorescent diamond was exhibited in the lecture, green by daylight, colourless by candle-light. It is mounted in the centre of an exhausted bulb (fig. 87), and the molecular discharge was directed on

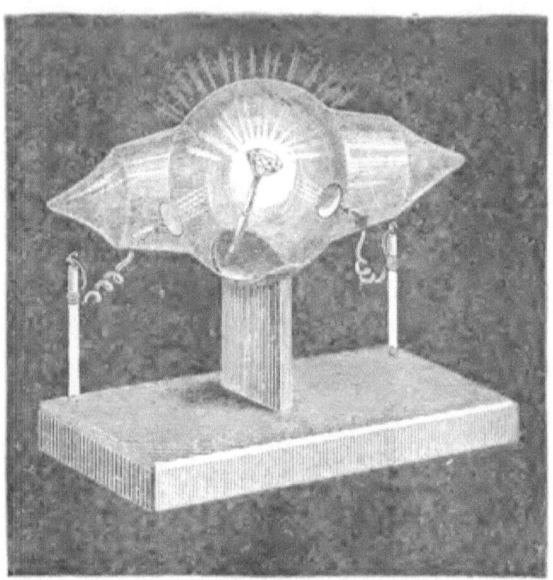

Fig. 87.

it from below upwards. On darkening the room, the diamond was seen to shine with as much light as a candle, phosphorescing of a bright green.

Next to the diamond the ruby is one of the most remarkable stones for phosphorescing. A tube shown in fig. 88 contained a fine collection of ruby pebbles. As soon as the induction spark was turned on, the rubies were seen to shine with a brilliant rich red tone, as if they were glowing hot. It scarcely matters what colour the ruby is, to begin with. In the tube of natural rubies there were stones of all colours—the deep red and also the pale pink ruby. There were some so pale as to be almost colourless, and some of the highly-prized tint of pigeon's blood; but under the impact of radiant matter they all phosphoresced with about the same colour.

Fig. 88.

Now the ruby is nothing but crystallized alumina with a little colouring matter. In another tube was some precipitated alumina prepared in the most careful manner. It had been heated to whiteness, and under the molecular discharge it glowed with the same rich red colour.

Fig. 89 represents a tube made to illustrate the depend-

Fig. 89.

ence of the phosphorescence of the glass on the degree of exhaustion. The two poles are at a and b, and at the end (c) is a small supplementary tube connected with the other by a narrow aperture, and containing solid caustic potash. The tube had been exhausted to a very high point, and the potash heated so as to drive off moisture and injure the vacuum. Exhaustion had then been recommenced, and the alternate heating and exhaustion repeated until the tube had been brought to the state in which it was exhibited. When the induction spark was first turned on, nothing was visible —the vacuum was so high that the tube was non-conducting. The potash was then warmed slightly, so as to liberate a trace of aqueous vapour. Instantly conduction commenced, and the green phosphorescence flashed out along the length of the tube. The heat was continued so as to drive off more gas from the potash. The green got fainter, and then a wave of cloudy luminosity swept over the tube, and stratifications appeared, which rapidly got narrower, until the spark passed along the tube in the form of a narrow purple line. The lamp was taken away, and the potash allowed to cool; as it cooled, the aqueous vapour, which the heat had driven off, was reabsorbed. The purple line broadened out, and broke up into fine stratifications; these got wider, and travelled towards the potash tube. Then a wave of green light appeared on the glass at the other end, sweeping on and driving the last pale stratification into the potash; and, lastly, the tube glowed over its whole length with the green phosphorescence. If the tube is left to itself for some time, the green grows fainter and the vacuum becomes non-conducting.

155. Radiant matter proceeds in straight lines. The radiant matter, whose impact on the glass causes an evolution of light, absolutely refuses to turn a corner. Fig. 90 represents a V-shaped tube, a pole being at each extremity. The pole at the right side (*a*) being negative, the whole of the right arm was flooded with green light, but at the bottom it stopped sharply and would not turn the corner to get into the left side. When the current was reversed, and the left pole made negative, the green changed to the left side, always following the negative pole, and leaving the positive side with scarcely any luminosity.

To produce the ordinary phenomena exhibited by vacuum tubes, it is customary, in order to bring out the striking contrasts of colour, to bend the tubes into very elaborate designs. The luminosity caused by the phosphorescence of the residual gas follows all the convolutions into which skilful glass-blowers can manage to twist the glass. The negative pole being at one end and the positive pole at the other, the luminous phenomena seem to depend more on the positive

Fig. 90.

than on the negative at the ordinary exhaustion hitherto used to get the best phenomena of vacuum tubes. But at a very high exhaustion the phenomena noticed in ordinary vacuum tubes, when the induction spark passes through them—an appearance of cloudy luminosity and of stratifications—disappear entirely. No cloud or fog whatever is seen in the body of the tube, and with such a vacuum as is used in these experiments, the only light observed is that from the phosphorescent surface of the glass. Fig. 91 represents two bulbs, alike in shape and position of poles, the only difference being that one is at an exhaustion equal to a few millimetres of mercury—such a moderate exhaustion as will give the ordinary luminous phenomena—whilst the other is exhausted to about the millionth of an atmosphere. First

Fig. 91.

the moderately exhausted bulb (A) was connected with the induction coil, and the pole at one side (*a*) being retained always negative, the positive wire was put successively to the other poles with which the bulb is furnished. It was seen that as the position of the positive pole was changed, the line of violet light joining the two poles changed, the electric current always choosing the shortest path between the two poles, and moving about the bulb as the position of the wires was altered.

This, then, is the kind of phenomenon we get in ordinary

exhaustions. The same experiment was then tried with a bulb (B) that was very highly exhausted, and, as before, the side pole (*a*) was made the negative, the top pole (*b*) being positive. The appearance seen was very widely different from that shown by the last bulb. The negative pole was in the form of a shallow cup. The molecular rays from the cup crossed in the centre of the bulb, and, thence diverging, fell on the opposite side and produced a circular patch of green phosphorescent light. The positive wire was removed from the top and connected first to the side pole (*c*), then to the bottom pole (*d*); but the green patch remained where it was at first, unchanged in position or intensity.

We have here another property of radiant matter. In the low vacuum, the position of the positive pole is of every importance, whilst in a high vacuum the position of the positive pole scarcely matters at all; the phenomena seem to depend entirely on the negative pole. If the negative pole points in the direction of the positive, all very well; but, if the negative pole is entirely in the opposite direction, it is of little consequence: the radiant matter darts all the same in a straight line from the negative.

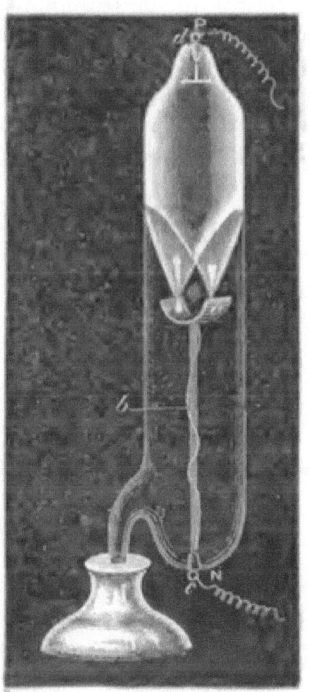

If, instead of a flat disc, a hemi-cylinder is used for the negative pole, the matter still radiates normal to its surface. The tube represented in fig. 92 illustrates this property. It contains, as a negative pole, a hemi-cylinder (*a*) of polished aluminium. This is connected with a fine copper wire, *b*, ending at the platinum terminal, *c*. At the upper end of the tube is another terminal, *d*. The induction-coil is connected so that the hemi-cylinder is negative and the upper pole positive, and, when exhausted to a sufficient extent, the projection of the molecular rays to a focus is very beautifully shown. The rays of matter, being driven from the hemi-cylinder in a direction normal to its surface, come to a focus and then diverge, tracing their path in brilliant green phosphorescence on the surface of the glass.

Another tube was shown, in which the focus of molecular rays was received on a phosphorescent screen instead of on the glass. The effect produced was most brilliant, and lighted up the whole table.

156. Radiant matter, when intercepted by solid matter, casts a shadow. Radiant matter comes from the pole in straight lines, and does not merely permeate all parts of the tube and fill it with light, as would be the case were the exhaustion less good. Where there is nothing in the way, the rays strike the screen and produce phosphorescence; and where solid matter intervenes, they are obstructed by it, and a shadow is thrown on the screen. In the pear-shaped bulb (fig. 93) the negative pole (a) is at

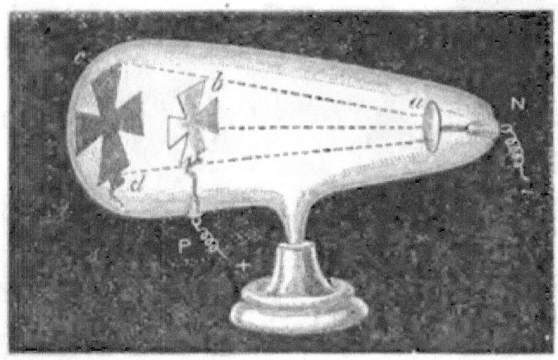

Fig. 93.

the pointed end. In the middle is a cross (b) cut out of sheet aluminium, so that the rays from the negative pole projected along the tube will be partly intercepted by the aluminium cross, and will project an image of it on the hemispherical end of the tube which is phosphorescent. When the coil was turned on, the black shadow of the cross was clearly seen on the luminous end of the bulb (*c, d*). Now, the radiant matter from the negative pole has been passing by the side of the aluminium cross to produce the shadow; the glass has been hammered and bombarded till it is appreciably warm, and at the same time another effect has been produced on the glass—its sensibility has been deadened. The glass has got tired, if the expression may be used, by the enforced phosphorescence. A change has been produced by this molecular bombardment which will prevent the glass from responding easily to additional excitement; but the part that the shadow has fallen on is not tired—it has not been phosphorescing at all, and is perfectly fresh; therefore, on the cross being thrown down* so as to allow the rays from the negative pole to fall uninterruptedly on to the end of the bulb, the black cross (*c, d*, fig. 94) was seen suddenly to change to a luminous one (*e, f*)

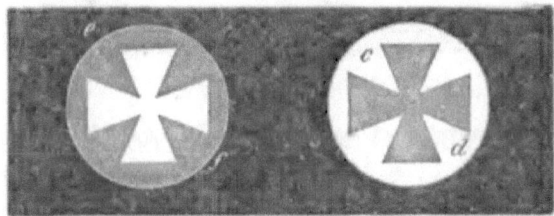

Fig. 94.

because the background was now only capable of faintly phosphorescing, whilst the part which had the black shadow on it retained its full phosphorescent power. The stencilled image of the luminous cross soon dies out. After a period of rest the glass partly recovers its power of phosphorescing, but it is never so good as it was at first.

Here, therefore, is another important property of radiant matter. It is projected with great velocity from the

---

* This could be done by giving the apparatus a slight jerk, the cross having been constructed with a hinge.

negative pole, and not only strikes the glass in such a way as to cause it to vibrate and become temporarily luminous while the discharge is going on, but the molecules hammer away with sufficient energy to produce a permanent impression upon the glass.

157. Radiant matter exerts strong mechanical action where it strikes. We have seen, from the sharpness of the molecular shadows, that radiant matter is arrested by solid matter placed in its path. If this solid body is easily moved, the impact of the molecules will reveal itself in strong mechanical action. Fig. 95 represents

Fig. 95.

an ingenious piece of apparatus, consisting of a highly-exhausted glass tube, having a little glass railway running along it from one end to the other. The axle of a small wheel revolves on the rails, the spokes of the wheel carrying wide mica paddles. At each end of the tube, and rather above the centre, is an aluminium pole, so that, whichever pole is made negative, the stream of radiant matter darts from it along the tube, and, striking the upper vanes of the little paddle-wheel, causes it to turn round and travel along the railway. By reversing the poles the wheel can be arrested and sent the reverse way; and if the tube be gently inclined, the force of impact is observed to be sufficient even to drive the wheel up-hill.

This experiment therefore shows that the molecular stream from the negative pole is able to move any light object in front of it.

The molecules being driven violently from the pole, there should be a recoil of the pole from the molecules, and by arranging an apparatus so as to have the negative pole movable and the body receiving the impact of the radiant matter fixed, this recoil can be rendered sensible. Fig. 96 represents an apparatus whose appearance is not unlike an ordinary radiometer, with aluminium discs for vanes, each disc coated on one side with a film of mica. The fly is supported by a hard steel instead of glass cup, and the needle-point on which it works is connected by means of a wire with a platinum terminal sealed into the glass. At the top of the radiometer bulb a second terminal is sealed in. The radiometer therefore can be connected with an induction coil, the movable fly being made the negative pole.

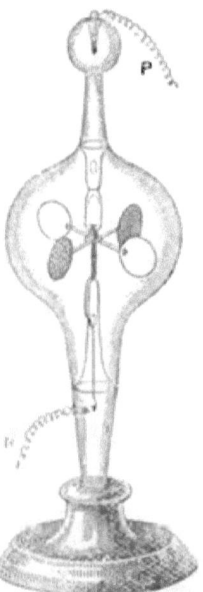

Fig. 96.

For these mechanical effects the exhaustion need not be so high as when phosphorescence is produced. The best pressure for this electrical radiometer is a little beyond that at which the dark space round the negative pole extends to the sides of the glass bulb. When the pressure is only a few millims. of mercury, on passing the induction current a halo of velvety violet light forms on the metallic side of the vanes, the mica side remaining dark. As the pressure diminishes, a dark space is seen to separate the violet halo from the metal. At a pressure of half a millim. this dark space extends to the glass, and rotation commences. On continuing the exhaustion the dark space further widens out and appears to flatten itself against the glass when the rotation becomes very rapid.

Fig. 97 represents another piece of apparatus which illustrates the mechanical force of the radiant matter from the negative pole. A stem (a) carries a needle-point in which revolves a light mica fly ($b\ b$). The fly consists of four square vanes of thin clear mica, supported on light

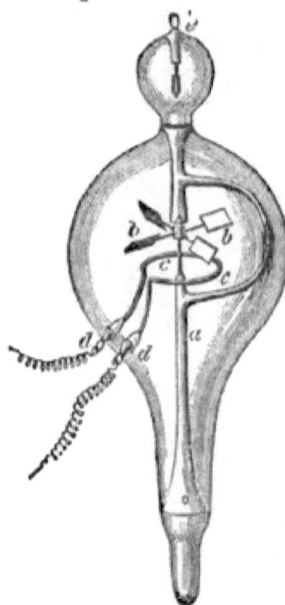

Fig. 97.

aluminium arms, and in the centre is a small glass cap which rests on the needle-point. The vanes are inclined at an angle of 45° to the horizontal plane. Below the fly is a ring of fine platinum wire ($c\ c$), the ends of which pass through the glass at $d\ d$. An aluminium terminal (e) is sealed in at the top of the tube, and the whole is exhausted to a very high point. Wires from the induction coil were attached, so that the platinum ring was made the negative pole, the aluminium wire (e) being positive. Instantly, owing to the projection of radiant matter from the platinum ring, the vanes rotated with extreme velocity. Thus far the apparatus had shown nothing more than the previous experiments had prepared

us to expect; but another phenomenon was then exhibited. The induction coil was disconnected altogether, and the two ends of the platinum wire connected with a small galvanic battery; this made the ring $c\ c$ red-hot, and under this influence it was seen that the vanes spun as fast as they did when the induction coil was at work.

Here, then, is another most important fact. Radiant matter in these high vacua is not only excited by the negative pole of an induction-coil, but a hot wire will set it in motion with force sufficient to drive round the sloping vanes.

158. Radiant matter is deflected by a magnet. We now pass to another property of radiant matter. Fig. 98 shows a long glass tube, very highly exhausted; it has a negative pole at one end (a) and a long phosphorescent screen (b, c) down the centre of the tube.

In front of the negative pole is a plate of mica (*b, d*) with a

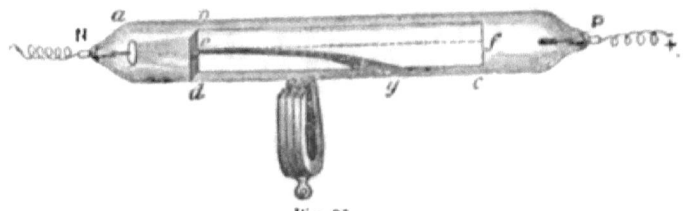

Fig. 98.

hole (*e*) in it; and when the current was turned on, a line of phosphorescent light (*e, f*) was projected along the whole length of the tube. A powerful horse-shoe magnet was now placed beneath the tube, and the line of light (*e, g*) became curved under the magnetic influence, waving about like a flexible wand as the magnet was moved to and fro.

This action of the magnet is very curious, and, if carefully followed up, will elucidate other properties of radiant matter. Fig. 99 represents a tube exactly similar, but having at

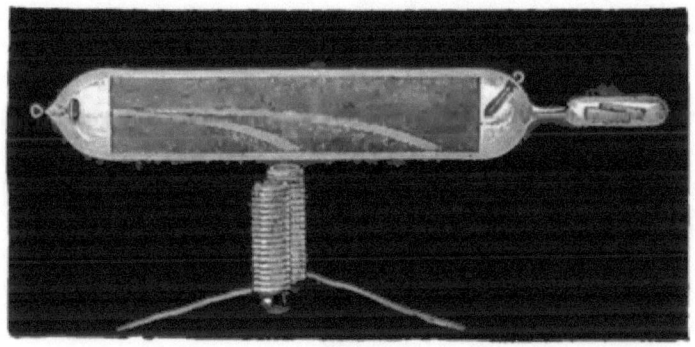

Fig. 99.

one end a small potash tube, which, if heated, will slightly injure the vacuum. When the induction current is turned on, the ray of radiant matter is seen tracing its trajectory in a curved line along the screen, under the influence of the horse-shoe magnet beneath. Let us observe the shape of the curve. The molecules shot from the negative pole may be likened to a discharge of iron bullets from a mitrailleuse, and the magnet beneath will represent the earth curving the trajectory of the shot by gravitation. The curved trajectory of the shot is accurately traced on the luminous screen. Now suppose the deflecting force to remain constant, the curve traced by the projectile varies with the velocity. If more powder be put in the gun, the velocity will be greater and the trajectory flatter; and if a denser resisting medium be interposed between the gun and the target, the velocity of the shot will be diminished, and it will move in a greater curve and come to the ground sooner. The velocity of this stream of radiant molecules cannot well be increased by strengthening the battery, but they can be made to suffer greater resistance in their flight from one end of the tube to the other. In the experiment shown, the caustic potash was heated with a spirit-lamp, and so a trace more gas was thrown in. Instantly the stream of radiant matter responded. Its velocity was impeded, the magnetism had longer time on which to act on the individual molecules, the trajectory became more and more curved until, instead of shooting nearly to the end of the tube, the "molecular bullets fell to the bottom before they had got more than half-way.

It is of great interest to ascertain whether the law governing the magnetic deflection of the trajectory of radiant matter is the same as has been found to hold good

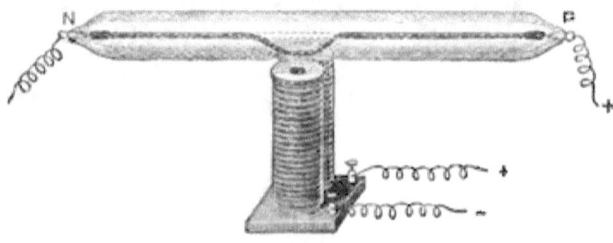

Fig. 100.

at a lower vacuum. The experiments just described were made with a very high vacuum. Fig. 100 represents a tube with a low vacuum. When the induction spark is turned on, it passes as a narrow line of violet light joining the two poles. Underneath is a powerful electro-magnet. On making contact with the magnet, the line of light dips in the centre towards the magnet. On reversing the poles, the line is driven up to the top of the tube. We notice the difference between the two phenomena. Here the action is temporary. The dip takes place under the magnetic influence; the line of discharge then rises and pursues its path to the positive pole. In the high exhaustion, however, after the stream of radiant matter had dipped to the magnet, it did not recover itself, but continued its path in the altered direction.

By means of the little wheel (fig. 101) Mr. Crookes was able

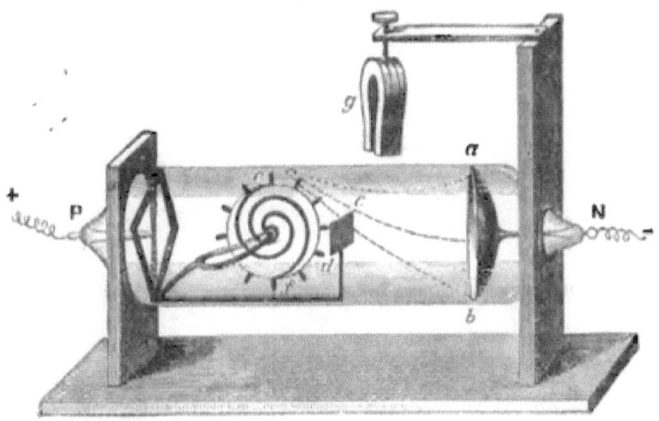

Fig. 101.

to show the magnetic deflection in the electric lantern. The negative pole (a, b) is in the form of a very shallow cup. In front of the cup is a mica screen (c, d), wide enough to intercept the radiant matter coming from the negative pole. Behind this screen is a mica wheel (e, f) with a series of vanes, making a sort of paddle-wheel. So arranged, the molecular rays from the pole a, b will be cut off from the wheel, and will not produce any movement. A magnet g was now put over the tube, so as to deflect the stream over or under the obstacle $c\ d$, and the result was rapid motion in one or the other direction, according to the way the magnet was turned. The image of the apparatus was thrown on the screen. The spiral lines painted on the wheel showed which way it turned. The magnet was arranged to draw the molecular stream so as to beat against the upper vanes, and the wheel revolved rapidly as if it were an over-shot water-wheel. On turning the magnet so as to drive the radiant matter underneath, the wheel slackened speed, stopped, and then began to rotate the other way, like an under-shot water-wheel. This reversal can be repeated as often as the position of the magnet is reversed.

We have mentioned that the molecules of the radiant matter discharged from the negative pole are negatively electrified. It is probable that their velocity is owing to the mutual repulsion between the similarly electrified pole and the molecules. In less high vacua, such as that shown in fig. 100, the discharge passes from one pole to another,

Fig. 102.

carrying an electric current as if it were a flexible wire. Now, it is of great interest to ascertain if the stream of radiant matter from the negative pole also carries a current. Fig. 102 is an apparatus which decides the question at once. The tube contains two negative terminals (*a*, *b*) close together at one end, and one positive terminal (*c*) at the other. This enables two streams of radiant matter to be sent side by side along the phosphorescent screen; or, by disconnecting one negative pole, only one stream.

If the streams of radiant matter carry an electric current, they will act like two parallel conducting wires and attract one another; but if they are simply built up of negatively electrified molecules, they will repel each other.

The upper negative pole (a) was first connected with the coil, and the ray was soon shooting along the line d, f. The lower negative pole (b) was then brought into play, and another line (e, h) darted along the screen. Instantly the first line sprung up from its first position, $d f$ to $d g$, showing that it was repelled, and the lower ray was also deflected downwards: *therefore the two parallel streams of radiant matter exerted mutual repulsion, acting not like current carriers, but merely as similarly electrified bodies.**

159. Radiant matter produces heat when its motion is arrested. Another property of radiant matter is that the glass gets very warm where the green phosphorescence is strongest. The molecular focus on the tube (fig. 91) is intensely hot.

We can render this focal heat more evident if we allow it to play on a piece of metal. The bulb (fig. 103) is furnished with a negative pole in the form of a cup (a). The rays will therefore be projected to a focus on a piece of iridio-platinum (b) supported in the centre of the bulb.

The induction-coil was first slightly turned on so as not to bring out its full power. The focus played on the metal, raising it to a white heat. By bringing a small magnet near, the focus of heat was deflected just as was the luminous focus in the other tube. By shifting the magnet the focus can be driven up and down, or drawn completely away from the metal, so as to leave it non-

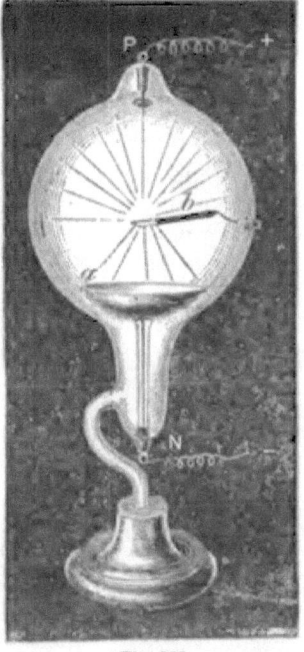

[Fig. 103.

* See page 202.

luminous. On withdrawing the magnet so as to let the molecules have full play again, the metal became white-hot. On increasing the intensity of the spark, the iridio-platinum glowed with almost insupportable brilliancy, and at last melted.*

Thus we see that electricity, whatever it may be, can so act on matter that when the molecules are free to obey its impulses they are projected in various directions with force sufficient to produce the bombardment effects which we have described. It is by investigations of this nature that we may hope some day to obtain a knowledge of what electricity is, and when we know that we shall be in a fair way to comprehend much of the hidden mechanism of nature.

Questions on Chapter XXIII.

1. What are the three states of matter? Give illustrations to show how they differ from each other.

2. What is meant by radiant matter?

3. What is the effect of radiant matter on phosphorescent bodies?

4. Describe an experiment to show that radiant matter proceeds in straight lines.

5. A vacuum tube has a cup-shaped negative terminal and a positive one not in the focus of the cup. Describe the nature of the discharge (a) when the vacuum is low, (b) when it is high.

6. Describe an experiment to show that radiant matter when intercepted by a screen casts a shadow.

7. Describe an experiment to show that radiant matter exerts mechanical force where it strikes.

8. Explain the difference between the deflection caused by a magnet of an ordinary discharge and a radiant one.

9. Describe an experiment to show that radiant matter produces heat when its motion is arrested.

* The highest vacuum Mr. Crookes has yet succeeded in obtaining has been the 1-20,000,000th of an atmosphere, a degree which may be better understood if we say that it corresponds to about the hundredth of an inch in a barometric column three miles high.

CHAPTER XXIV.

ELECTROSTATIC INDUCTION.

160. When two bodies at different electric pressures are connected by a wire, a current will, as we know, flow from that at high pressure to that at low; but when they are separated from each other by an insulator, a new series of phenomena occur, which we have now to consider.

Suppose we have some insulated conductors, such as metal balls hanging by silk threads, and we connect some of them to the positive pole of a battery of a great many cells, and some to the negative. The balls will become "charged" at the pressures of the battery; and when disconnected from the battery, they will still retain their charges. The same "charging" may be done by other means than by the use of a battery. A positive charge may be produced by briskly rubbing a glass rod with a piece of silk, and a negative one by rubbing a stick of sealing-wax.

In these experiments, an apparatus called an electric-machine (fig. 104) is used, which consists of a glass plate, which is turned by a handle, and pressed on by a cushion. It is provided with a conductor, to collect the charge. It is more convenient and cheaper than a large battery for experiments when we wish to study the effects of electric pressure only, and not current, as it produces very high pressures (tens of thousands of volts), but indefinitely small currents, perhaps not a ten-thousandth part of an ampère.

Fig. 104.

161. Actions between charged bodies. Charged bodies attract and repel each other according to the following laws :—*

Positively charged bodies repel each other.
Negatively charged bodies repel each other.
Positively charged bodies attract negatively charged ones ; and,
Negatively charged bodies attract positively charged ones.

Or, generally, *electric charges of the same kind repel each other ; of different kinds, attract.*

162. Law of force. *Electrified bodies attract or repel with a force varying inversely as the square of the distance,*

Fig. 105.

* The attractions and repulsions being very small, these different forces may be conveniently observed by a balance apparatus, such as that in fig. 105, where the rod can be laid in a little wire cradle, suspended by a silk thread,

i.e. if at 1 inch distance the force is unity, at 2 inches it would be $\frac{1}{4}$, at 3 inches $\frac{1}{9}$, &c.

163. Induced charges. When a charged body is placed near an insulated conductor, a charge opposite to its own is induced on the near side of the conductor; and an equal charge, similar to its own, on the further side. A charged body will therefore attract light bodies not previously charged. The laws of these actions are similar to those of the action of magnets on each other and on soft iron.

Now these actions are of no great scientific importance in themselves; but it is of great importance to study the means by which they are propagated through the intervening medium.

Mechanical forces are propagated by means of a *strain*, or distortion of the connecting medium. For instance: if a weight is to be moved by pulling a rope, the molecules of the rope nearest to the puller are slightly separated from the next, and these latter, in returning to their original distance from the first, pull the next after them, and so on, until the force is transmitted to the weight.

The problem, then, that we have before us is: Given the known experimental facts which we have just been considering; given that there is an action, which we call induction, across air and other insulators from an electrified body to other bodies in the neighbourhood; that the induction causes these attractions, and repulsions, and "inducings" of electrifications which we have spoken of, what is the machinery by means of which this induction acts? What is the nature of the lever, the rope, or the pushing pole, which strains, and pulls, and pushes across the air, or glass, or other non-conductor which we place between the induced and inducing bodies? We must attempt to answer this question bit by bit, and our first attempt shall be based on the difference between induction and conduction.

When a piece of glass or other insulator is placed in contact with the conductor of an electric machine, it is thrown into a state of strain and distortion, for it is "charged," but the electricity does not escape through it,

When, however, a metal or other conductor takes the place of the glass, there is no appearance of such a state of strain at all. What is the explanation of this? It is this. Equally in conductors and insulators a state of strain occurs, but in conductors *this state of strain is continually giving way,* while in insulators it does not do so. To keep up the state of strain in a conductor would be as difficult as to keep up a pressure of steam in a boiler with a large hole it.

Let us consider a mechanical experiment in illustration—only in illustration, remember, not in explanation—of what I mean. Fig. 106 represents a vessel, U, connected to the water-pipes at one end, and to a pressure-gauge, S, at the other. There is no escape for the water, it cannot flow or move, and the gauge shows a considerable pressure. Turning on the tap T, will allow a stream of water to escape. The pressure and strain is relieved and the gauge falls; that is, as soon as the state of constraint gives way and the current flows, it is seen that the strain no longer exists. In the analogous electrical case, bodies in which the state of constraint easily gives way do not show the phenomena of strain or induction, but allow the electricity to flow freely, and these are called conductors; while, on the other hand, bodies which have a great power of resistance to the straining force can be greatly strained without allowing a current of electricity to flow. These are called insulators or non-conductors. When such a body is subjected to a powerful straining or inducing electric force, it exhibits the phenomena of strain or induction very strongly.

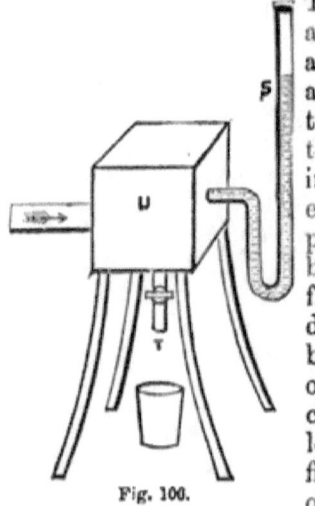
Fig. 106.

164. The particles of glass move more freely over each other when hot than when cold, and hence we should expect that hot glass would yield more easily to a straining force than cold glass would. The following experiment shows that this is the case. Fig. 107 represents a glass flask containing mercury, and set in a dish of mercury. The mercury inside is connected to the electric machine, and that outside to the earth. On working the machine, it is found, first, that no electricity can escape through the flask; se-

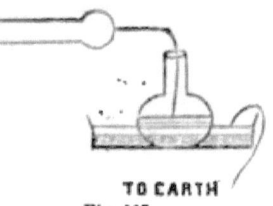
Fig. 107.

condly, that there is a strong induced charge on the mercury outside. Now let the mercury be made hot. It heats the glass, the particles move more freely over each other, the glass yields to the straining force, electricity escapes through it, and at the same time all induction ceases.

We have already spoken of condensers, which are made in various forms. The form of condenser most suitable for our present inquiry is the old form, known as the Leyden jar.

165. The Leyden jar, in its most common form, consists of a wide-mouthed bottle, coated inside and out with tinfoil. The wooden stopper supports a brass knob, which communicates, by means of a wire or chain, with the inside coating. In order that the inside and outside coatings may be well insulated from each other, they do not reach quite to the top of the jar. Thus the jar forms a system of two conductors (the tinfoils), separated by a thin insulator (the glass). We connect the outer tinfoil to earth. If we connect the knob of the jar to the machine, and work the latter, we can charge the inside tinfoil positively, and, on removing the machine, this tinfoil will retain its charge for a considerable time, as is shown by the electroscope E on the knob. (Fig. 108.)

This insulated electrified conductor now acts by induction through the glass of the jar, and induces negative electricity on the outer tinfoil conductor.

Here, then, we have our two conductors oppositely charged, acting on each other inductively through the glass.

Fig. 108.

166. Some idea of the intensity of the strain may be obtained by "discharging" the jar, as it is called. This can be done by a pair of what is called "discharging tongs," which consist of a conveniently-shaped conductor fixed to an insulating handle. We hold the "tongs" so that one knob touches the outer conductor of the Leyden jar (fig. 109), and then bring the other knob of the "tongs" near the knob of the jar, which, we remember, is connected to the inner coating.

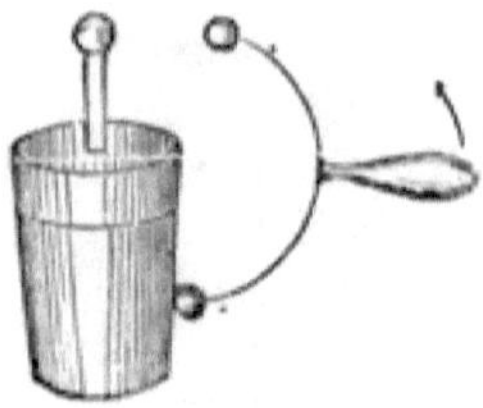

The strain between the conductors is now taking place through two different insulators; that is, first through the glass of the jar, second, through the air between the two knobs, viz., the knob of the jar and the upper knob of the tongs. The glass is strong enough to resist the straining force of such a charge of electricity as the jar now has. We now bring the knobs nearer together. The straining force across the air between them gets greater and greater, and, at the same time, as the thickness of the air diminishes, its power of resistance, or of sustaining the state of strain, gets less and less, till at last, the air breaks and gives way, and the electricities rush together with a flash and a report.

Fig. 109.

167. Residual charge. The straining force of the charge which we gave to the inner coating is removed with the charge, and immediately after the flash and discharge the falling of the electroscope shows that there is no electricity whatever on either coating of the jar. But, in a very short time, the electricity seems to be returning; a slight motion of the electroscope ball shows that a slight charge has returned to the inner coating. On applying the discharging tongs a spark occurs as before, only it is very much feebler, and the jar is now completely discharged.

168. Mechanical illustration. What does this mean? Where did the second charge of electricity come from? Let us consider a mechanical experiment, which will help us to an explanation. We use a strip of gutta-percha, of which the lower end is fixed to a block. We now bend it down with the finger, and suddenly let it go. It flies up nearly, but not quite to the vertical position, rests an instant, and then moves slowly on till it is quite vertical.

If a spring index had been applied to it, it would have been seen that while pressed down it exercised a strong upward pressure. At the moment when it was at rest a little way from the vertical it would be exercising no pressure, and then it would be seen that as it began to again move towards the vertical, it would again exercise pressure. The gutta-percha was **strained** or distorted by the finger. When the straining force was removed, the strain, suddenly, nearly disappeared, but not quite. Then, in the course of the next few **minutes**, the disappearance of the strain or distortion was completed slowly.

The electrical case is exactly analogous. The pressure of the finger represents the first charging of the inner tinfoil; the straining of the gutta-percha represents the electrical straining of the glass. The pressure on the finger by the strained india-rubber represents the induction on the outer conductor. As in the gutta-percha, when the straining force is removed, the strain or distortion nearly disappears, and the upward pressure exercised by it entirely ceases, so in the Leyden jar, when the inducing electricity is taken away, the **strain** of the glass almost vanishes, and the induced charge disappears.

The strain or distortion of the glass, however, has only almost, but not entirely, disappeared; **and now** that there is no straining **force** interfering, **the** particles of the glass move over each **other** slowly, and in the course of **a** few minutes **return to their** normal state. But now, **while** the **inner** conductor has remained insulated, a change has occurred in the electrical arrangement of the particles of glass adjoining it. The state of strain has altered. They have changed from a more **to a** less distorted shape.

Now, in the ordinary phenomena of induction, what happens **when** we alter the **state** of strain of an insulator by bringing a charged body near it? Why, **it** induces electricity on any adjoining conductor. Similarly in the present **case,** when the elasticity of the glass brings it from a more to a less strained state, and so alters the state of strain, a charge is produced on the insulated conductor, and this is the residual charge which we have been inquiring about.

169. Effect of tapping the glass. We must notice that this residual charge returns slowly and gradually. Now, when a body is mechanically distorted, and is returning to its normal state by virtue of its elasticity, anything which enables the particles to move more freely over each other, such as tapping or jarring, will hasten that return. If, for instance, we have a heaped up tray of sand slowly returning to its normal unstrained state of being level under the action of gravitation, any tapping of the tray will hasten the recovery from the state of strain; that is, hasten the return of the surface to a level state by enabling the particles to slide more freely over each other.

Now, if our supposition that these induction phenomena are the effects of strain, and that the residual charge is the returning of the distorted particles of glass to their normal state, is correct, any tapping or jarring of the glass should hasten this return; that is, hasten the appearance of the residual charge. In the *Phil. Trans.* for 1876, Dr. Hopkinson has shown that this actually occurs, and I have repeated his experiment in the following form at the Royal Institution.

For this purpose, as we wish to measure electrification, we must use an electrometer. The instrument which was used is called the quadrant-electrometer, and is the invention of Sir William Thomson.

Fig. 110.

The right-hand half of fig. 110 is a diagram of the essential parts of it. A sort of brass pill-box, supported horizontally, is cut into four quarters or quadrants, each of which is insulated from the one next it, but connected to the one diagonally opposite to it. An aluminium needle, N N, is suspended, so that it can swing like a compass needle inside the pill-box. The needle has a strong positive charge. When wires from the inner and outer coatings of a Leyden jar are connected to the wires a and b respectively, so that the unshaded quadrants are positive and the shaded ones negative, it will be seen that the action of all four quadrants is to turn the needle in the direction of the hands of a watch.

The motion of the index needle itself is very small, but attached to it is a small mirror. In the lecture, light from a lime-light fell on it, and was reflected on to the screen, forming a bright spot. The least motion of the needle and mirror of course moved the light-spot along the screen. The amount of motion was noted by means of the scale attached to the screen.

The Leyden jar in this case was made in a form somewhat different to that which we have been considering. The insulator was, as before, a small glass bottle,* but the conductors consisted of strong sulphuric acid. Some was put inside the jar, and some in a glass dish in which the jar was set. To charge the jar a platinum wire came from the electric machine to the acid inside the bottle, while that outside was connected to earth.

The jar was charged for two or three minutes and then discharged. The inner and outer coatings were then connected for about a minute by holding the wires from them together between the finger and thumb. The wires were then separated and connected to the quadrants of the electrometer, the earth connection being removed. The spot of light showing the motion of the needle at once began to move along the scale at the rate of about three inches per second, showing that the residual charge was coming slowly out.

On tapping the edge of the bottle briskly with a piece of

* A bottle, about four inches high, intended for making a small Leyden jar, was used.

P

hard wood, the pace at which the light-spot moved was at once trebled,* showing that while tapping is going on, the residual charge returns much more quickly.

If it is wished to repeat the experiment, we can discharge the electrometer, and bring the light-spot back to zero by holding the wires together for a moment between the finger and thumb.

Thus we see that any mechanical vibration communicated to the particles of glass increases their freedom of motion among each other, and, therefore, enables them to recover more quickly after they have been thrown into a state of strain by electric induction.

This experiment is nearly conclusive, I think, as to induction being a state of strain, and this concludes what we have to say about the Leyden jar.

170. How strain is propagated. We will now begin to inquire into the nature of this strain, and try to find out a little about how it is propagated from place to place.

Is it propagated simply in straight lines? does the inducing electricity stretch out an arm through the insulator, and pull at the second conductor? or, does it act only on the particles of the insulator which are nearest to it, leaving it to them to act on the next set, and so to carry on the strain from particle to particle till it arrives at the second conductor?

Faraday asked himself the question, and it occurred to him that there was a very simple method of arriving at an answer. If the induction is propagated from particle to particle of the insulator, it can travel along any direction where there is a continuous chain of insulating particles, whether this chain forms a straight line or not; in other words, it can *turn a corner*. If it were a " direct action at a distance " (whatever that may mean) it could only travel in straight lines.

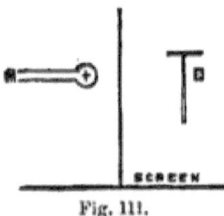
Fig. 111.

The following experiment is a modification of one designed by Faraday to show that induction can take place in curved

* This experiment was plainly seen by a large audience.

lines. No induction can take place through a metal screen which is connected to earth. The simplest way to prove this is by experiment. Fig. 111 represents a large metal screen connected to earth; the electric machine being placed on one side of it, and a gold-leaf electroscope* on the other. However strongly the machine was worked there was no divergence of the leaves.

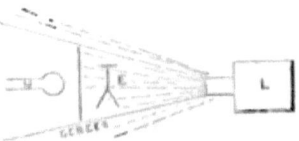

Fig. 112.
L = Light. E = Electroscope.
M = Machine.

I then (fig. 112) took away this screen, and put in a smaller one. In order to be sure that this, though smaller, was still large enough to cut off all straight lines from every part of the machine to the electroscope, I put the lime-light L as shown in fig. 112. It was seen that the optical shadow of the electroscope fell entirely on the screen, and the shadow of the screen entirely covered the machine. On working the machine the leaves diverged widely.

How did the induction get to them? Our experiment with the large screen shows that it could not have passed through the screen; that with the lime-light shows that it could not have come in a *straight* line past the edge of the small screen; and, therefore, we see that it must have come in *curved* lines round the edge of the small screen.

This experiment shows a point of difference between the projection through air of light and of electric induction, for while the edge of the optical shadow is almost on the straight line from the source of light through the edge of the screen, the electrical shadow does not extend nearly so far, but the induction curves considerably round the edge of the interposing screen, and extends in every direction, in which there is a continuous chain of insulating matter.

171. Induction precedes discharge. I think the experiment which we have just described proves the existence of induction in curved lines. It is possible, however, actually to show a curved line of induction. If I connect an electroscope to a knob placed near an electric

---

* This is a very delicate apparatus for detecting small electrical charges.

machine and connected to earth, when we work the machine so that sparks pass slowly, the electroscope shows a strong induced charge, which increases to a maximum just before each spark and immediately after the spark falls to zero. (Fig. 113.)

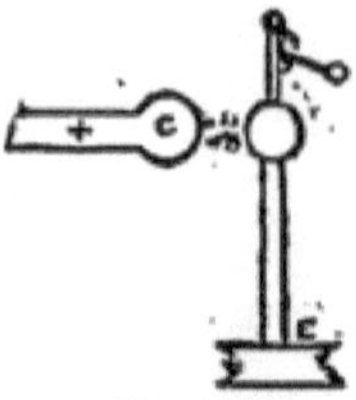

Fig. 113.

This experiment shows that induction precedes discharge. All that we know about the subject shows that this is a universal law, namely, that there must always be induction along the whole path between the conductors before discharge can take place. It is clear that this law ought to hold, for discharge is only the sudden breaking down of a state of strain, and there can be no breaking down of strain except where strain exists, and induction is strain.

The fact of a spark passing along any path shows that induction was previously taking place along that path. It does not, however, show that the whole induction was along that particular path even a very small fraction of a second before the discharge. The induction might have been, and probably was taking place along many paths. When, however, the insulator broke down at the weakest point, and the spark began to pass, the whole of the induction at once transferred itself to the line of discharge as being the path offering least resistance, and then the breaking down and relief of the strain was completed along that path.

Given, then, that induction precedes discharge, if we see

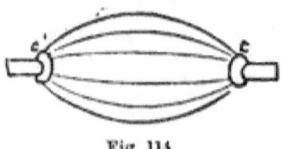

Fig. 114.

curved discharges, we shall know that there was previously curved induction. Discharge often consists of a barrel-shaped bundle of sparks. (Fig. 114.) They are, in fact, the curved lines of force, or lines of induction, or lines of strain, produced in visible shape. The centre lines are straight, and the strongest induction takes place along them; but induction strong enough to produce discharge takes place in curved lines through all the particles on all sides of the centre line.

172. Lines of force are real. These lines of force are

real, and, I may almost say, tangible things. They can be attracted and shaped by the hand and other conductors. If I place my knuckle near the lines, they bend out towards it. This means that the positively electrified particles of air induce negative electricity on my hand, and then the two electricities attract each other, and displace the whole line of force. It would be difficult to conceive the possibility of attracting an "action at a distance."

We have shown that the induction is a state of strain, and we have studied the direction of the strain. We now ask, what is its nature? Faraday showed experimentally that the lines of force attracted each other, so that, if a number of them were side by side forming a "bundle," their mutual attraction drew them together as if the bundle had been tied up more tightly.

Maxwell has since pointed out that this is what occurs whenever a rope is mechanically stretched. The pull tending to lengthen the rope is accompanied by a pressure tending to make the rope thinner. To show this *lateral* pressure, we may use an india-rubber tube (fig. 115). When we stretch it the sides press on whatever is inside it. Whenever a mechanical tension occurs it is accompanied by a pressure at right angles to it.

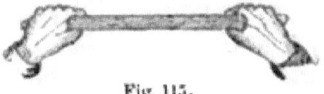

Fig. 115.

Maxwell has shown mathematically that an electric induction acting as a tension along the lines of force is always accompanied by an exactly equal pressure at right angles to them. Electric induction, or tension, is a tension of exactly the same kind as the tension of a rope, and the medium which can support a certain induction force before breaking and allowing a spark to pass may be said to have a certain strength in exactly the same sense as a rope may be said to have a certain strength. Sir William Thomson has found that the electric strength of air at ordinary temperature and pressure is 9600 grains per square foot.

Finally, Mr. De la Rue has actually seen in one of his vacuum tubes a star of light showing a rain of particles thrown off at right angles to the main discharge.

Questions on Chapter XXIV.

1. Describe the actions between insulated electrified bodies. State when attraction occurs, when repulsion.

2. A body A is positively charged, and one B negatively. Do they attract or repel? What happens when the charges of both are reversed? What when A is only reversed, what when B?

3. State the law according to which the attractions and repulsions vary with the distance.

4. When a piece of glass or other insulator is placed in contact with the conductor of an electric machine, it is thrown into a state of strain and distortion, for it is "charged," but the electricity does not escape through it. When, however, a metal or other conductor takes the place of the glass, there is no appearance of such a state of strain at all. What is the explanation of this?

5. Describe a mechanical experiment to illustrate your answer?

6. Describe an experiment to show that the particles of glass yield more freely to electric strain when the glass is hot than when it is cold.

7. Describe the Leyden jar.

8. Describe the process of discharging a Leyden jar.

9. What is "residual charge"? Explain it.

10. Give a mechanical illustration of residual charge.

11. Describe an experiment to show that the return of residual charge can be hastened by a mechanical vibration of the particles of the glass.

12. Describe an experiment to show how electric strain is propagated. Can it turn a corner?

13. Describe an experiment to show curved lines of induction.

CHAPTER XXV.

SPECIFIC INDUCTIVE CAPACITY.

We have seen that Induction is transmitted from particle to particle of insulators, and that its phenomena are exhibitions not of some direct action passing through the insulator, but of something actually existing in the particles of the insulator itself; that it is in some peculiar straining of these particles that the causes of the phenomena will be found.

173. Specific inductive capacity. One of the first questions which now presents itself is, Do all insulators on which a given inducing charge acts suffer an equal strain, and therefore exhibit the same quantity of inductive action at the other side, or, on the contrary, does the same charge of electricity strain different insulators differently and produce induced changes of different strengths; in other words, are there in different insulators different capacities of receiving strain from a given straining charge—differences of specific "strainability"—that is, differences of *Specific Inductive Capacity?*

Various experiments satisfied Faraday that the latter is the case, and that different bodies have different specific inductive capacities.

174. Definition. First, however, let us make it quite clear what is meant by the term specific inductive capacity. Let a certain charge of electricity be acting inductively across air upon a neighbouring conductor, and let the sizes of the conductors and the distance between them be such that the strength of the charge induced on the second conductor is equal to unity. Let the whole space between the conductors be now filled with some other insulator. The strength of the induced charge will now be no longer unity, but it will have some other value. The number which represents this value is called the "specific inductive capacity" of the substance between the conductors; in other words, the specific inductive capacity of a substance is the ratio of the inductive action across it to that across air. Air being taken as the standard, its specific inductive capacity is called unity.

We will now examine some of the various methods by which, from time to time, the specific inductive capacities of various bodies have been measured.

175. Faraday's experiments. As Faraday was the discoverer of specific inductive capacity, we will begin with his experiments. Faraday's wish was to construct a Leyden jar, of which the metallic coatings should be fixed, and always in the same relative positions, while the insulator should be movable; so that various Leyden jars could be set up, which should be exactly alike in all respects, except in the nature of their insulator, which could be made to consist either of air, glass, sulphur, or any other substance.

If such jars could be constructed, and if differences were observed in their behaviour, these differences could only be due to differences of induction through the different insulators, or to differences of specific inductive capacity.

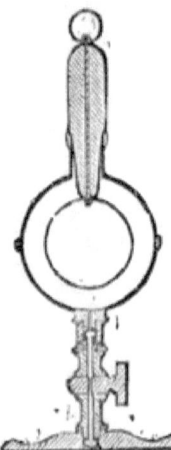

Fig. 116.

The apparatus consists of a metal ball (fig. 116), which can be surrounded by a larger hollow one. The outer ball is made in two pieces, so as to allow the inner one to be placed inside it. There is a space of 0·62 inch between the surfaces of the balls. The inner ball is supported by an insulating stem of shellac passing through a hole in the outer one. A wire which passes up inside the shellac allows the inner ball to be put in connection with an electric machine, electrometer, or other apparatus.

The space between the balls contains the insulator. It

may be **air, or the** whole or **part** of the space may be filled with glass, sulphur, &c. Faraday preferred only to fill half the space, and then to calculate what the effect would have been if the whole space had been filled. He, therefore, prepared his insulators in the form of hemispherical cups.

He constructed two of these Leyden jars, so that he could observe simultaneously their actions with different insulators, and endeavoured to make them precisely alike. If they had not been precisely alike, there would have been **a** difference in their behaviour which would have been due, not to difference in the specific inductive capacities **of** the insulators, but to differences in the shape **and** size **of** the jars. In order to make sure that they were exactly alike, he made an elaborate series of preliminary experiments. We need not go into all the details of these preliminary experiments, but we can indicate the principle of them in a few words.

If the **two** "jars" are equal they will have equal capacities for electricity. That is, under similar circumstances they will each hold an equal quantity of electricity.

To determine this, Faraday first charged one apparatus only, and measured the charge. He then connected the **two** together, so that the charge **divided** itself between them. He **then** separated them and remeasured the charge of the first one. If the **second apparatus had** the **same** capacity as the first, it **would have taken away** exactly half the charge. If it had **a greater or less** capacity, it would **have** taken more or less than half the charge.

The "jars" were, therefore, adjusted till the charge left in each after division with the second was exactly one-half of the original charge before division. I have said that "the amount of charge was measured," but have not yet explained how that was done. Here again as the electro-*scopes* which we have been experimenting with only show the existence or non-existence of a charge, but do not measure its amount, we require an electro*meter*. The electrometer used by Faraday to measure the induction was the invention of Coulomb, and is called the "torsion balance." Descriptions of it will be found in all books on physics; **but, as** it **is now** obsolete, we will not trouble ourselves with an **account of it.**

Having adjusted his torsion balance and determined the equality of his two Leyden jars, Faraday was ready to commence his measurements of specific inductive capacity. He kept one apparatus full of air, and placed in the other a hemispherical cup of shellac. He then compared the inductive actions through the two machines, and he at once found that the induction through the shellac apparatus was greater than that through the air apparatus in the proportion of 176 to 113, or 1·55 to 1. In this case the air apparatus had been charged first.

Another set of experiments in which the shellac apparatus was charged first gave a ratio of 1·37 to 1. This difference, which is considerable, is accounted for by the fact that the experiment takes some time, and that there is a constant leakage of electricity going on; and in the one set the effect of the leakage would be to give too high a result, and in the other to give too low a one. The mean, therefore, will not be far from the truth.

Faraday gives as his result that the induction through the apparatus half-filled with shellac is 1·5 times that through the one full of air. From this he calculates that the ratio of the specific inductive capacity of shellac is to that of air rather more than 2 to 1.

I have purposely avoided attempting to give the exact details of Faraday's method of working, for two reasons. One is that it is an exceedingly difficult thing to understand, as the inductive actions through the different insulators are compared by an indirect method, to follow which requires a considerable familiarity with the laws of induction; and the other is that I thought it unnecessary to burden your memories with the minor details of a method of working, which, owing to the invention of improved apparatus, no experimenter would now adopt, wonderful as it was in its day, and wonderful for all time as *are* the results which were obtained by it.

Faraday continued his experiments on other substances, and the following is a general table of his results: Shellac, 2; sulphur, 2·24; flint glass, 1·76 or more.

After Faraday came numerous experimenters, who have published results more or less accordant for the specific inductive capacities of many insulators. Space will not admit of my giving you an account of all the methods which have been used.

176. Gordon's experiments. As an illustration of modern methods we will examine the outlines of some determinations which I made some few years ago. We may here mention that insulators across which induction takes place are called "dielectrics," from the Greek διά = across.

The great difficulty which all investigators of specific inductive capacity have met with has been due to the fact that, if a dielectric is charged for any appreciable time, some of the charge is "absorbed," and the phenomena of "residual charge" complicate the observations.

In the experiments we are about to describe the effects of absorption are guarded against in two ways.

(1) The electrified metal plates of the condenser do not touch the dielectrics; and

(2) The charging only lasts $\frac{1}{12000}$ of a second.

177. The induction balance. Fig. 117. The chief instrument used is a very complicated condenser called the "Induction Balance," the general plan of which is due to Sir Wm. Thomson and Professor Clerk Maxwell.

It consists essentially of five circular parallel metal discs, $a\ b\ c\ d\ e$; $b\ c\ d\ e$ are fixed, and a can be moved parallel to itself by means of a screw. $a\ c\ e$ are 6 inches in diameter, $b\ d$ are 4 inches. There is a space of about 1 inch between each plate and the one next it.

The source of electrification (coil poles in fig. 117) is one that gives equal and opposite pressures. Where the double sign ($\pm \mp$) is given, it means that the sign of the electrifications can be rapidly reversed. We will for the present consider the electrifications of all parts of the apparatus to have the upper sign, and the reversing engine not to be at work.

We see that one pole of the coil is connected to the outside plates a and e, and the other pole to the centre plate c.

The two small plates, b and d, are connected to the quadrants of an electrometer.*

The centre plate c can never produce any deflection of the needle, because, being placed half-way between the

* See page 208.

220   School Electricity.

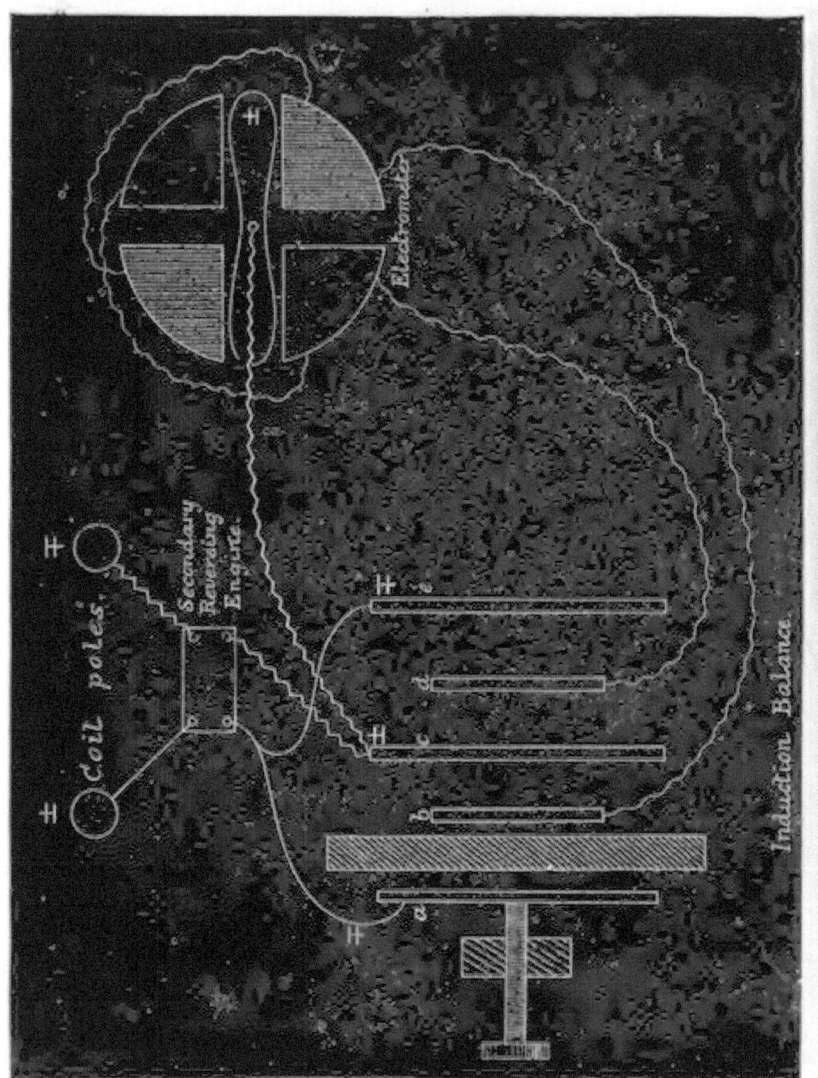

Fig. 117.

small plates, it will produce an equal and similar charge on each of them, and therefore equal and similar charges on the quadrants.

The outer plates will also produce no effect on the needle as long as there is only air in the balance and they are placed symmetrically—that is, as long as distance *a b* is equal to distance *c d*.

If, however, *a* is moved by its screw away from *b* (so as to make distance *a b* greater than *c d*), there will be a *less* inductive action from *a* to *b* than from *c* to *d*, and the needle will be deflected in the direction which shows that the unshaded quadrants are most strongly electrified.

But if, on the other hand, any dielectric of greater specific inductive capacity than air be placed between *a* and *b*, there will be a *greater* inductive action from *a* to *b* than from *c* to *d*, and the needle will be deflected in the opposite direction.

It is clear, then, that if we insert a dielectric, and at the same time screw *a* away from *b*, we can find a position for *a* where the increased induction through the dielectric is exactly balanced by the decrease due to the greater distance, and then the needle will remain at zero.

The distance which *a* will have to be moved to compensate any given dielectric plate will depend only on the thickness of the latter, *and on its specific inductive capacity*.

In the experiments we read the position of *a*, which brings the electrometer needle to zero; 1st, when there is only air in the balance; 2nd, when the dielectric is inserted.

The difference of these two readings is the distance which *a* has been moved. We measure the thickness of the dielectric, and then we can calculate the specific inductive capacity by the following mathematical formula.

178. The formula of calculation. Let the reading of *a*, when there is only air in the balance, be a_1, and that when the dielectric is inserted, a_2; then $(a_2 - a_1)$ is the distance which *a* has had to be moved.

A dielectric plate of thickness *b*, and specific inductive capacity K, acts like a plate of air of thickness $\frac{b}{K}$. That is to say, the capacity of a condenser whose dielectric plate had a thickness b, and specific inductive capacity K, would be equal to that of a similar condenser having, for its dielectric, a plate of air of thickness $\frac{b}{K}$.

We must remember that when we insert a dielectric plate of thickness b into the balance, we displace a plate of air of the same thickness b.

The effect then of inserting the dielectric is to *increase* the capacity of the condenser, consisting of the plates a and b, as much as if we had brought the plates nearer by a distance b, and further apart by a distance $\frac{b}{K}$; that is (as b is greater than $\frac{b}{K}$), as if we diminished the distance between the plates by a quantity $b - \frac{b}{K}$.

But as we have moved a so as to keep the needle at zero, we have produced an exactly equal *decrease* of capacity by increasing the distance between a and b by a quantity $(a_2 - a_1)$. It is clear that this real increase of distance must be exactly equal to the imaginary decrease of distance produced by the dielectric, and we must have

$$b - \frac{b}{K} = (a_2 - a_1);$$

or, in other words,

$$K = \frac{b}{b - (a_2 - a_1)}$$

and this formula was used to calculate the results of the experiments.

We note that we do not require to know the distances $a_1 b$ or $a_2 b$, but only their difference, which is much more easily measured.

179. The reversals. We have hitherto supposed the needle to be charged positively in the ordinary way. Let us suppose the equilibrium not to be established, but the action of a to be greater than that of e, and the electrifications to have the upper signs. a and e will induce $(+)$ electricity on b and d, and $(-)$ on all four quadrants; but the electrification of the shaded quadrants will be the strongest, and the needle will turn to the right (in the direction of the hands of a watch).

Now, suppose the electrifications of a c e to be all reversed in sign, but to have the same numerical values as before; a and e will now induce a $(-)$ charge on b and d, and a $(+)$ one on the quadrants. The shaded quadrants would still be the strongest, but the needle would now turn to the *left*. It is clear that if, as in actual work, the reversals were very rapid, the needle would merely receive rapidly alternating impulses in the two directions, and no disturbance of the equilibrium would produce any deflection.

To escape from this difficulty, it has been arranged that the needle, instead of being permanently charged, should be connected to the plate c.

The sign of the charge of the needle is then reversed with the reversals of the charges of the balance.

Let now the electrifications have the upper signs, and the shaded quadrants be strongest. The quadrants will be $(-)$ and the needle $(+)$, and, the force being attractive, the needle will turn to the right.

Now let the electrification be reversed, the quadrants will be $(+)$ and the needle $(-)$. The force will still be attractive, and the deflection in the same direction as before.

In practice, when the electrifications of the five plates, the dielectric, the needle, and the four quadrants were all being reversed 12,000 times per second, the deflection of the needle was perfectly steady, and exactly under the control of the screw of a.

A motion of a of $\frac{1}{1000}$ inch usually moved the spot of light on the electrometer scale about one millim.

The rapid reversals (12,000 per second) were obtained by means of an induction coil and a special high-speed break.

A secondary reversing engine was used to again reverse the electrifications on their way to the balance about 30 times per second, in case there should be any preponderance of either $(+)$ or $(-)$ after the first reversal.

For details of the apparatus and method of working the reader is referred to my "Electricity," 2nd ed. vol. i. p. 109,

180. Results. The following is a general table of the results obtained for different substances:—

Name of Dielectric.	Specific Inductive Capacity.
Glass—	
Double extra-dense Flint	3·164.
Extra-dense Flint	3·054.
Dense Flint	..
Light Flint	3·013.
Hard Crown	3·108.
Common Plate, two specimens—	
No. 1	3·258 ⎫ mean 3·243.
,, 2	3·228 ⎭
Ebonite, four slabs—	
No. 1	2·2697 ⎫
,, 2	2·2482 ⎬ mean 2·284.
,, 3	2·3097 ⎪
,, 4	2·3077 ⎭
Best quality Gutta Percha	2·462.
Chatterton's compound	2·547.
India-rubber—	
Black	2·220.
Grey vulcanized	2·497.
Solid Paraffin, specific gravity at 11° C. = ·9109, melting point 68° C. Six slabs cut in planing machine; results corrected for cavities—	
No. 1	1·9940 ⎫
,, 2	1·9784 ⎪
,, 3	1·9969 ⎬ 1·9936.
,, 4	2·0126 ⎪
,, 5	1·9654 ⎪
,, 6	2·0143 ⎭
Shellac	2·74.
Sulphur	2·58.
Bisulphide of Carbon	1·81.

Specific Inductive Capacity.

The importance of these experiments lies in their bearing on the theory which connects the phenomena of electric induction with those of light, which we shall discuss later.

Professor Boltzmann and Messrs. Ayrton and Perry have, by means of a special apparatus, measured the specific inductive capacities of various gases with the following results:—

Gas.	Specific Inductive Capacity.	
	At 0° C. and 760 millims.	At 15°–17° C. and 760 millims.
Air at 0° C, and 760 millims. taken as 1·000000.		
Vacuum ·999410.		
Carbonic Acid	1·000356	1·000302
Hydrogen	·999674	·999660
Carbonic Oxide	1·000100	1·000060
Nitrous Gas (NO)	1·000394	1·000348
Olefiant Gas	1·000722	{ 1·000618 { 1·000576
Marsh Gas	1·000354	1·000300

Ayrton and Perry.

Dielectric.	Specific Inductive Capacity.
Air	1·0000
Vacuum	·9985
Carbonic dioxide	1·0008
Hydrogen	·9998
Coal gas	1·0004
Sulphuric dioxide	1·0037

Questions on Chapter XXV.

1. Define specific inductive capacity.
2. Describe Faraday's method of measuring it.
3. What is a "dielectric"?

4. Describe fully Gordon's method of measuring specific inductive capacity.

5. What does the fact that different bodies have different specific inductive capacities teach us as to the way in which electric induction is propagated?

CHAPTER XXVI.*

ANALOGIES BETWEEN THE MUTUAL ACTIONS OF CURRENTS, OF MAGNETS, AND OF ELECTRIFIED BODIES, AND THOSE OF BODIES PULSATING AND VIBRATING IN FLUIDS.

181. Professor Bjerknes' researches. At the Paris Electrical Exhibition of 1881, Professor C. A. Bjerknes, of Christiania, exhibited some most remarkable experiments, showing that it is possible to imitate most of the well-known actions between currents and magnets by means of bodies pulsating and vibrating in water and other fluids. Their importance lies in the fact that they afford a possible

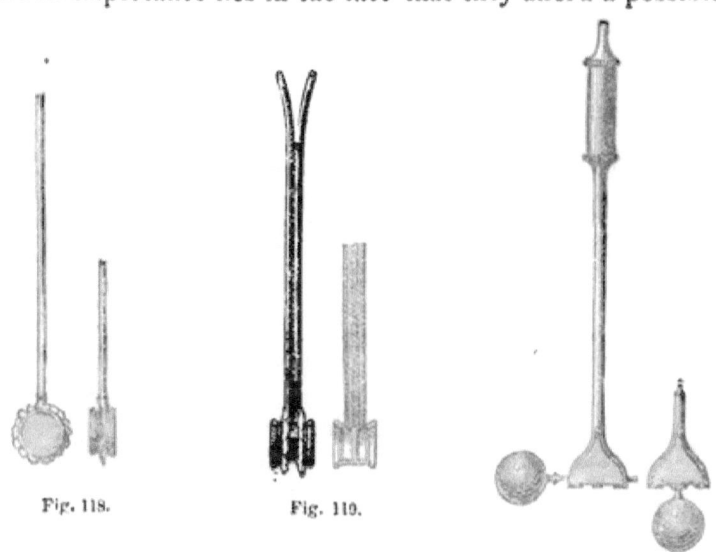

Fig. 118.   Fig. 119.

Fig. 120.

* See my "Electricity," 2nd ed. vol. ii. p. 36.

clue to the nature of the mechanism which transmits electric and magnetic forces through space.

The apparatus used consists of a glass trough, filled with water, in which the vibrating and pulsating bodies are placed. These latter consist of little drums, with elastic india-rubber ends, and which are made to pulsate by drawing air in and out by means of pumps. These pumps are without valves, but are constructed like an ordinary child's squirt, so that the air is drawn in and out at each motion of the piston. There are two pumps which actuate the two pulsating bodies whose mutual actions it is desired to study. They are driven rapidly by a hand-wheel, and by altering the position of one of the crank pins, they can be made to work either in the same or in opposite phases.

In the experiments, one pulsating body is held in the hand, and the other pivoted on a stand, so that it is free to move like a compass needle under the attraction or repulsion of the first one.

Fig. 118 shows one of the pulsating drums.

Fig. 119 shows another, which has a central diaphragm and two tubes, so that its two faces can, if desired, be made to pulsate in opposite phases.

Fig. 120 shows the arrange-

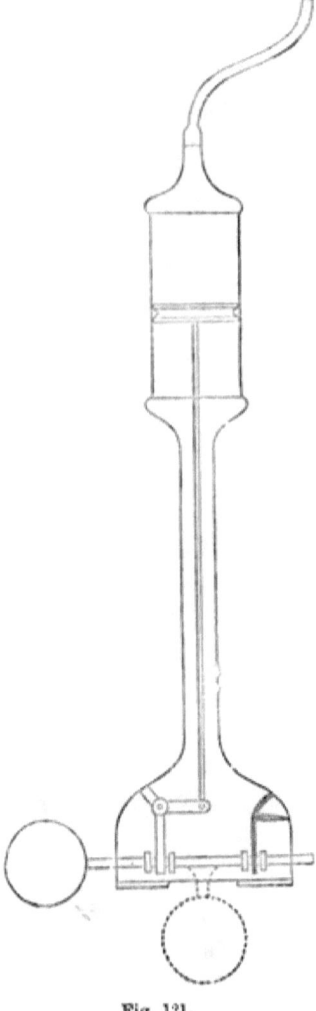

Fig. 121.

ment for a vibrating body, where the ball moves backward and forward along a horizontal line. It is moved by the mechanism shown in fig. 121.

182. Phases. Two pulsating bodies are said to be in the same phase when they both expand or both contract at the same instant, and in opposite phases when one expands while the other contracts.

The diagrams, pp. 230, 231, illustrate this analogy between pulsating and vibrating bodies, and magnetic poles and electrified bodies. It is found that there is a close analogy between the mutual actions of pulsating and vibrating bodies, and of magnetic poles and electrified bodies, but that in all cases the analogy is *inverse,* as may be seen on opposite page.

The force in all four cases varies *inversely as the square of the distance between the attracting and repelling bodies.*

183. Attraction of light bodies and of soft iron. If the suspended body is disconnected from the pump, then a pulsating body will repel it in the same way as an electrified body attracts light objects, or a magnet attracts soft iron.

184. Attraction or repulsion of compass needle. A body oscillating along a horizontal axis, which is free to turn, will follow or fly from a pulsating or vibrating body just as a compass needle will follow or fly from the pole of a magnet. The direction of the force depends on the phase, as explained in **182**.

185. Two magnets of unequal strength. If two magnetic poles of the same polarity, but of which one is much stronger than the other, are placed a little distance apart, they will repel, but if brought *near* together, they *attract,* as the large one induces in the small one a polarity opposite to its own and stronger than its natural polarity.

Similarly, two pulsating bodies moving in the phase which produces attraction, and one of which is much larger than the other, will attract when they are a little distance apart, but *repel* when they are *near.*

186. Diamagnetism. Faraday suggested that many of the phenomena of diamagnetism may be accounted for by

ELECTRIFIED BODIES.	MAGNETIC POLES.	PULSATING BODIES.
Two bodies *similarly* electrified *repel* each other.	Two magnetic poles of the *same* name *repel* each other.	Two bodies pulsating in the *same* phase, i.e. posed faces of the drums are moving in opposite the same instant (figs. 122, 123), *attract* each

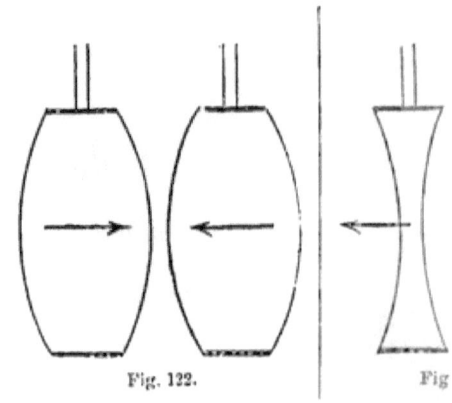

Fig. 122.   Fig

| Two bodies *oppositely* electrified *attract* each other. | Two magnetic poles of *opposite* name *attract* each other. | Two bodies pulsating in *opposite* phase, i.e. posed faces of the drums are moving in the same the same instant (figs. 126, 127), *repel* each |

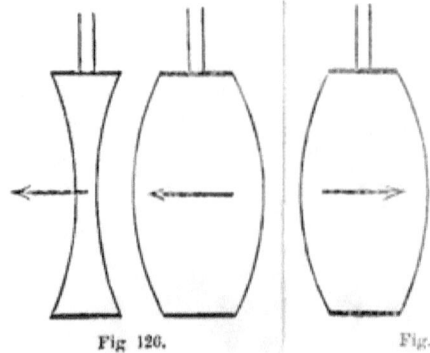

Fig 126.   Fig.

OSCILLATING BODIES.

Two oscillating bodies moving in *opposite* directions at the **same** instant (figs. 124, 125), *attract* each other.

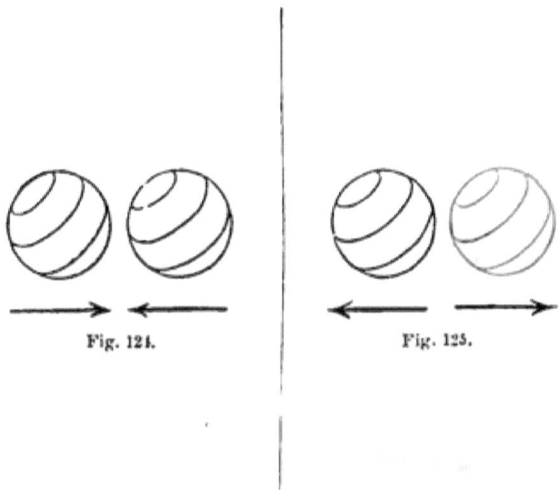

Fig. 124. Fig. 125.

Two oscillating bodies moving in the *same* direction at the same instant (figs. 128, 129) *repel* each other.

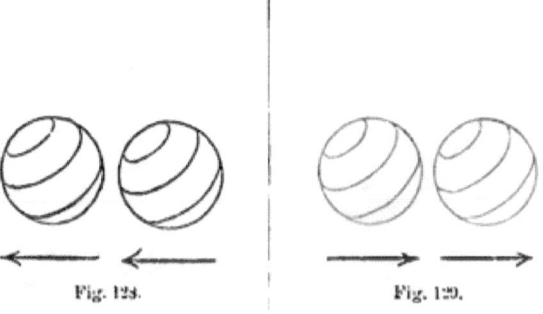

Fig. 128. Fig. 129.

supposing all bodies to be paramagnetic, but of different strengths, and that the apparent repulsion observed with bismuth and other bodies is only due to the stronger attraction exercised on the medium in which they are immersed. It is probable, however, that this explanation is not sufficient to account for all the phenomena observed. The analogous

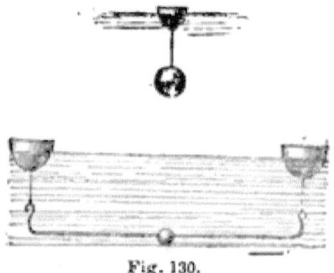

Fig. 130.

Fig. 131.

case in the Bjerknes' experiments is that the actions on a body are opposite, according to whether it is lighter or heavier than the medium in which it is immersed.

Bodies *heavier* than water (fig. 130) are *attracted* by a pulsating body; bodies lighter than water (fig. 131) are repelled.

Another way of arranging the same experiment is shown in fig. 132.

In each case we consider the body *heavier* than water as a type of a *diamagnetic* body, remembering that all the phenomena are inverse; that lighter than water to be similarly the (inverse) type of a paramagnetic body. Thus

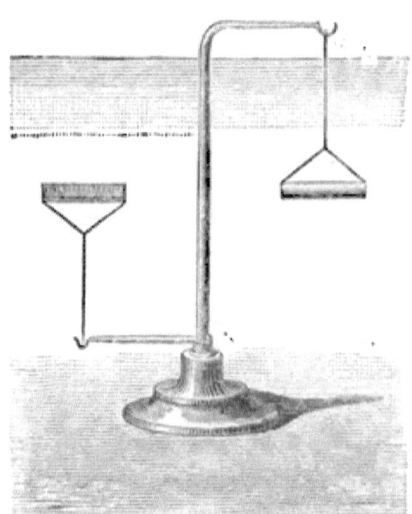

Fig. 132.

the body heavier than water is acted on like soft iron, the one lighter like bismuth.

If two magnetic poles of the same name be placed a little way apart, a piece of *iron* will be *repelled* from between them; if they are of opposite names, it will be drawn in.

Similarly, if two drums are arranged as in fig. 133, a body *lighter* than water will be *attracted* to the centre if the drums vibrate in the same phases, and will be repelled if they vibrate in the opposite phases.

With bismuth, and with a body heavier than water, these effects will be reversed.

187. Lines of force. By means of the apparatus shown in fig. 134, Professor Bjerknes has succeeded in tracing out the *lines of force* in the water due to the various pulsating and vibrating bodies experimented on, in a form which has enabled him to compare them with the corresponding lines produced by magnets and currents as traced by iron filings, in the manner described on page 14.

The apparatus shown in fig. 126 consists of a heavy metal ball, supported on a stand by means of a light steel spring. This being placed at various points, always oscillates along the direction of the line of force at that point, i.e. along the direction of the wave in the water. A fine rod attached to the top of it projects from the water, and carries a camel's-hair brush, which records the direction of vibration on the under side of a piece of smoked glass.

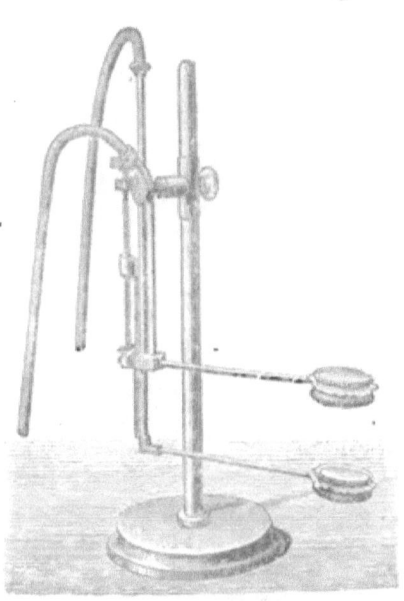

Fig. 133.

A series of curves have been obtained, which show that

Fig. 131.

the lines of force between vibrating and pulsating bodies in water are exactly similar in form to those between converging electric, magnetic, and electro-magnetic forces.

188. Mr. Stroh has since repeated Professor Bjerknes' experiments, and has reproduced nearly all the phenomena by means of sound-waves in air—i.e. he has caused air, vibrating by sound, to transmit the forces in the same way as Professor Bjerknes transmitted them through water.

By these researches we see opened a possibility of explaining some of the mysterious mechanism of electric and magnetic attractions, without the necessity of supposing any force to be at work other than those with which common experience makes us familiar, for we see them all reproduced by vibrations of material fluids, which differ from our supposed ether only in the superior elasticity and smaller density of the latter, that is, they differ from it only in degree and not in kind.

Questions on Chapter XXVI.

1. Describe the experimental methods used by Professor Bjerknes to illustrate the analogy between electric and magnetic attractions, and the forces between bodies pulsating and vibrating in water.

2. **When** are pulsating bodies said to be in the same phase, when in opposite phases?

3. When are vibrating bodies said to be in the same phase, when in opposite phases?

4. Describe Bjerknes' experiment to illustrate the attraction of soft iron by a magnet.

5. Describe one to illustrate Diamagnetism.

6. Describe Bjerknes' method of tracing lines of force in water.

CHAPTER XXVII.

ACTION OF MAGNETISM AND ELECTRICITY ON POLARIZED LIGHT.*

189. Polarized light. Ordinary light consists of vibrations taking place always in planes at right angles to the direction of the ray, but in all directions in those planes. That is, if the ray travels along the axle of a wheel, the vibrations composing it are all in the plane of the wheel, but are executed along any or all of the spokes.

The effect of reflecting light at certain angles from certain substances, or of passing it through certain crystalline substances, is to cause all the vibrations to take place in the same direction—that is, along one spoke of the wheel and the spoke opposite to it.

The light is then said to be polarized. Now if the wheel, without being rotated, be slid along the axle, the spoke along which the vibrations take place will trace out a plane.

When no rotative force is applied to the polarized light, the vibrations all take place in this plane, and the light is said to be " plane-polarized."

If we twist the reflector or crystal, which we use as a polarizer, round the direction of the ray as axis, we shall shift this plane in the same way as if we cause the wheel to

---

* The student who is ignorant of the elements of the theory of polarized light is recommended to read "The Polarization of Light," by the late Wm. Spottiswoode, Pres. R.S., &c. Nature Series (Macmillan).

turn on its axis and so shift the spoke along which the vibrations take place; but, when the wheel is slid along the axis, this spoke will still trace out a plane, only that plane will not be the same as before. That is, if we turn the polarizing mirror or crystal, we turn the plane of polarization, but the light still remains plane-polarized.

We cannot detect by the eye in what plane light is polarized, or indeed whether or not it is polarized at all. In order to do so we have to take advantage of the following natural law:—

Transparent bodies which have the power of polarizing light in any given plane are opaque to light already polarized in a plane at right angles to that plane; and reflecting surfaces which have the power of polarizing light in a given plane will not reflect light which, when it falls on them, is already polarized in a plane at right angles to that plane.

Thus, to determine in what plane light is polarized, we have only to take a crystal which has the power of polarizing light in a certain plane fixed with regard to its axis, and to turn it round till the light is extinguished.

We then know that the light is polarized in a plane at right angles to that plane in the crystal.

190. Natural rotation. Certain natural substances, such as oil of turpentine, creosote, &c., possess the power of "rotating the plane of polarization;" that is, if a horizontal beam of light, polarized in a horizontal plane, be sent through a tube filled with one of these substances, its plane of polarization, on emergence, will no longer be horizontal, but will be inclined on one side or the other of the horizontal, according to the nature of the substance.

The amount of inclination for the same substance depends on the length of the substance travelled through.

191. Faraday's discovery of magnetic rotation. Faraday discovered that if a piece of a particular kind of glass known as "heavy glass" is placed between the poles of an electro-magnet, and a ray of polarized light sent through it from pole to pole, the plane of polarization is rotated. That is, if the light be supposed to go in, in a horizontal plane, it emerges, still plane-polarized, in a plane inclined to the horizontal. The direction of the inclination depends on the polarity of the magnet; if this be reversed, by reversing the current, the direction of the rotation is changed. Faraday afterwards found that many other substances could produce the same effect when magnetized.

Thus far the magnetic phenomenon appears to be analogous to the phenomenon of rotation by oil of turpentine, &c. There is, however, a very important difference between the two cases.

192. Difference between magnetic and natural rotation. If, instead of permitting the light to emerge, we let it fall perpendicularly on a mirror placed at the end of the column of turpentine or glass, and be reflected back, we shall find that, *in the case where the rotation is produced by the magnetic force, the amount of rotation is doubled; while in the case where the rotation is produced by the natural power of the substance, the rotation is annulled;* and further, if by silvering both ends of the heavy glass, or of the tube containing the oil of turpentine, we cause the light to pass backward and forward any number of times (fig. 135), we

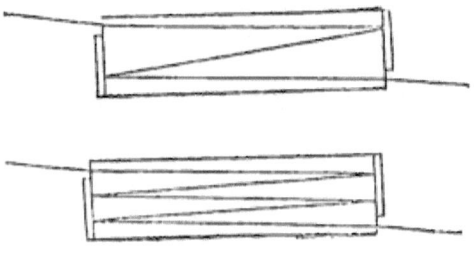

Fig. 135.

shall find that, in the case of the magnetic rotation, the rotation is equal to as many times the original rotation as the light passes through the substance; while in the case of the natural rotation, it is zero or equal to the original rotation, according to whether the light passes through the substance an even or an odd number of times.

A mechanical illustration may assist us to understand

this. Let us, as before, represent the plane of polarization by the direction of the spoke of a wheel, and let us slide the wheel backward and forward on the axis, and cause it to rotate at the same time so that the spoke is always in the plane of polarization at the point where the wheel is on the axis.

The case of natural rotation will then be illustrated by considering the turning of the wheel to be produced by the guiding action of a spiral or long screw thread cut on the axis. Thus, as the wheel moves along, the spoke traces out a twisted surface, while, when the wheel is slid back again, the spoke comes back along the same surface.

The case of magnetic rotation, however, is illustrated by considering the wheel *to be turned constantly in the same direction by some external force*, such as the pulling of a string wound round its circumference while the wheel is slid backward and forward. Let us suppose as before that the spoke always lies in the plane of polarization; the twisted surface traced out by it will now be different, as, on the wheel beginning to slide back along the axis, the spoke will not return on its old path, but will trace out a new surface whose twist is equal and opposite to the twist of the first surface.

193. Faraday's paper. Faraday communicated his discovery to the Royal Society on November 6th, 1845, in a paper entitled,—

"On the Magnetization of Light and the Illumination of Magnetic Lines of Force : i. Action of Magnets on Light; ii. Action of Electric Currents on Light; iii. General Considerations."

He commences his paper as follows :—

194. "Action of magnets on light. I have long held an opinion, almost amounting to conviction, in common, I believe, with many other lovers of natural knowledge, that the various forms under which the forces of matter are made manifest have one common origin ; or, in other words, are so directly related and mutually dependent, that they are convertible, as it were, one into another, and possess equivalents of power in their action. In modern times, the proofs of their convertibility have been accumulated to a very considerable extent, and a commencement made of the determination of their equivalent forces.

"This strong persuasion extended to the powers of light, and led, on a former occasion, to many exertions, having for their object the discovery of the direct relation of light and electricity, and their mutual action in bodies subject jointly to their power; but the results were negative and were afterwards confirmed in that respect by Wartmann.

"These ineffectual exertions, and many others which were never published, could not remove my strong persuasion derived from philosophical considerations; and, therefore, I recently resumed the inquiry by experiment in a most strict and searching manner, and have at last succeeded in *magnetizing and electrifying a ray of light, and in illuminating a magnetic line of force.*"

In the course of his paper, Faraday showed that the same effect as is produced by a magnet can be produced on a suitable transparent body placed inside a helix, the ray of light travelling along the axis of the helix. Fig. 136 shows

Fig. 136.

the method of arranging the experiment to illustrate the rotation in a liquid (bisulphide of carbon), to an audience.

195. Direction of the rotation. The plane of polarization is turned in the same direction as that in which the current flows when the substance is diamagnetic, in the opposite direction when it is paramagnetic.

196. General conclusions. The following are some of Faraday's general conclusions:—

"Thus is established, I think, for the first time, a true direct relation and dependence between light and the magnetic and electric forces; and thus a great addition made to the facts and considerations which tend to prove **that all natural forces are tied together, and have one common origin.** It is, no doubt, difficult in the present state **of our** knowledge to express our expectation in exact terms; **and** though I have said that another of the powers of nature is, in these experiments, directly related to the rest, I ought, perhaps, rather to say that another form of the great power is distinctly and directly related to the other forms; or **that** the great power manifested by particular phenomena in particular forms **is here** further identified and recognized by the direct relation **of** its form **of light to** its forms of electricity and magnetism.

"The relation existing between *polarized* light and magnetism **and** electricity is even more interesting than if it had been shown to exist with common light only. It cannot but extend to common light; and, as it belongs to light made, **in a** certain respect, more precise in its character and **properties by** polarization, it collates and connects it with **these** powers, in that duality **of** character which they possess, and yields an opening, which before was wanting to **us** for the **appliance** of these powers **to the** investigation of the nature of this and other **radiant agencies.**

"The magnetic **forces do not act on** the ray of light **directly and without the intervention of** matter, but through **the mediation of the substance in which** they **and** the ray **have a** simultaneous **existence;** the substances and the **forces giving to and** receiving from **each other the power of acting on the light.** This is shown by **the non-action of a vacuum or** of **air or** gases; and **it is** also further shown by the special degree **in** which different matters possess the property. That magnetic force **acts upon** the ray **of** light always with **the** same character of manner and in **the** same direction * **independent of** the different varieties **of** substance, or their **states of solid** or liquid, or their **specific** rotative force, **shows that the** magnetic force **and the light have a** direct relation; **but that** substances are **necessary,**

* It has since been shown by Verdet that this is not the case. Paramagnetic substances rotate the light in a direction opposite to diamagnetics. See **my** "Electricity," vol. ii. p. **225.**

R

and that these act in different degrees, shows that the magnetism and the light act on each other through the intervention of the matter.

"Recognizing or perceiving *matter* only by its powers, and knowing nothing of any imaginary nucleus, abstract from the idea of these powers, the phenomena described in this paper much strengthen my inclination to trust in the views I have on a former occasion advanced in reference to its nature.*

"It cannot be doubted that the magnetic forces act upon and affect the internal constitution of the diamagnetic just as freely in the dark as when a ray of light is passing through it; though the phenomena produced by light seem, as yet, to present the only means of observing this constitution and the change. Further, any such change as this must belong to opaque bodies, such as wood, stone, and metal; for as diamagnetics, there is no distinction between them and those which are transparent. The degree of transparency can at the utmost, in this respect, only make a distinction between the individuals of a class.

"If the magnetic forces had made these bodies magnets, we could by light have examined a transparent magnet; and that would have been a great help to our investigation of the forces of matter. But it does not make them magnets, and therefore the molecular condition of these two bodies, when in the state described, must be specifically distinct from that of magnetized iron, or other such matter, and must be *a new magnetic condition;* and as the condition is a state of tension (manifested by its instantaneous return to the normal state when the magnetic induction is removed), so the *force* which the matter in this state possesses, and its mode of action, must be to us a *new magnetic force* or *mode of action* of matter.

"For it is impossible, I think, to observe and see the action of magnetic forces, rising in intensity upon a piece of heavy glass or a tube of water, without also perceiving that the latter acquire properties which are not only *new* to the substance, but are also in subjection to very definite

* "Exp. Res.," vol. ii. p. 284; or Phil. Mag., 1844, vol. xxiv. p. 136.

and precise laws, and are equivalent in proportion to the magnetic forces producing them."

197. Later experiments. Numerous experiments have been made with more powerful and accurate materials than were at Faraday's disposal, and the relative and absolute amounts of rotation caused by given forces on different substances have been determined.

The rotation has even been observed and measured in gases.

198. Kerr's discovery. Dr. Kerr has discovered that if strong *electric pressures* be applied to the two sides of a piece of glass, that they actually distort or strain its structure sufficiently to make it act on light no longer as a homogeneous body, but as a crystal.

Figs. 137 and 138 show the experiment arranged for lecture purposes.

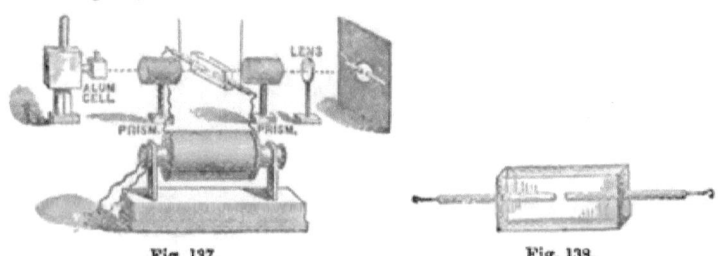

Fig. 137.   Fig. 138.

Two metal poles are let into the glass till they are within about three-sixteenths of an inch of each other. On the wires being connected to the poles of an induction coil, the strain sets up a temporary crystalline structure, which is shown by the action of the glass on the ray of polarized light. The alum cell is to stop the heat of the lamp.

QUESTIONS ON CHAPTER XXVII.

1. What is meant by "plane-polarized light"?
2. Describe the phenomenon of natural rotation of polarized light.

3. Describe the phenomenon of magnetic rotation of polarized light.

4. What is the difference between the two rotations? Give a mechanical illustration of it.

5. Describe a lecture experiment to show magnetic rotation by a helix.

6. What is the law governing the direction of rotation?

7. Describe Dr. Kerr's experiment for producing an artificial crystalline structure in glass by electro-static strain.

CHAPTER XXVIII.

CLERK MAXWELL'S ELECTRO-MAGNETIC THEORY OF LIGHT.*

199. ELECTRIC induction is a strain of some kind; and, when electric induction passes through space in which there is not any ordinary matter, we agree to call the unknown something that fills the space and transmits the strain an "ether."

Light is a strain of some kind; and when light passes through space where there is not any ordinary matter, we agree to call the unknown something that fills the space and transmits the strain an "ether."

All men of science are agreed that light consists of vibrations of an ether or very thin fluid which fills all space, and probably permeates all bodies.

200. Professor Clerk Maxwell's theory is briefly this:—

Electro-magnetic induction is propagated through space by strains or vibrations of the same ether which conveys the light vibrations, or, in other words, "light itself is an electro-magnetic disturbance."

Let us examine the evidence which causes us to believe that the luminiferous and the electro-magnetic ethers are one and the same.

The first point of resemblance between the modes of propagation of light and of electro-magnetic induction is that in both cases it can be shown mathematically that the disturbance is at right angles to the direction of propagation.

It is known that the waves of light take place in directions at right angles to the ray.

Professor Clerk Maxwell has shown that the directions

---

* See my "Electricity," 2nd. Ed. vol. ii. p. 305.

of both the magnetic and electric disturbances are also at right angles to the line of force.*

Fig. 139 shows Professor Maxwell's conception of a line of electric force.

The vertical line is the direction of the force, and the magnetic and electric disturbances are at right angles to it.

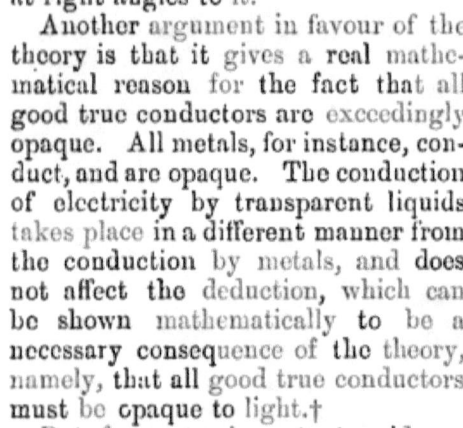

Fig. 139.

Another argument in favour of the theory is that it gives a real mathematical reason for the fact that all good true conductors are exceedingly opaque. All metals, for instance, conduct, and are opaque. The conduction of electricity by transparent liquids takes place in a different manner from the conduction by metals, and does not affect the deduction, which can be shown mathematically to be a necessary consequence of the theory, namely, that all good true conductors must be opaque to light.†

But far more important evidence in favour of the view that the ethers are not two, but one, is obtained by comparing the velocities with which optical and electro-magnetic disturbances are propagated under different circumstances.

If it can be shown that the velocity of electro-magnetic induction is sensibly the same as that of light, not only in air and vacuum, but in all transparent bodies, we shall be quite sure that there are not two ethers, but one; for it would be unreasonable to suppose that the whole of every part of space is filled with two ethers which are identical in the only properties which we can examine, but which are yet different and not the same.

And, further, if the velocities nearly agree, but not quite,

---

* They are also at right angles to each other.

† It must, however, be confessed that gold, silver, and platinum, when made into very thin plates, are not nearly so opaque as they should be according to the theory.

we must reserve our judgment; but we may be allowed to speculate on the possibility of the same ether vibrating somewhat differently when disturbed by electricity and by light.

201. Comparison of velocities in air and vacuum. The velocity of light has been measured experimentally in many ways.

The most recent experiments are those made by Professor Cornu,* in 1874, who found that in vacuo—

$$v = 3.004 \times 10^{10} \text{ centims. per second.}$$

The following are the results of older observations:—

Fizeau	3.14×10^{10}
Astronomical observations .	3.08 ,,
Foucault	2.98 ,,
Mean . . .	3.06 ,,

M. Cornu's experiments are, however, so greatly superior in accuracy to any of the older ones that we shall adopt his value, namely 3·004.

Now the refractive index of air is

$$1.000294.$$

The velocity of light in air is then:—

$$\frac{3.004 \times 10^{10}}{1.000294} = 3.0031 \times 10^{10}$$

Now the mean value † of the most recent determinations of the velocity of electro-magnetic induction in air is

$$v = 2.9857 \times 10^{10}$$

We may therefore say that *the velocities in air of light and of electro-magnetic induction are sensibly equal.*

202. Velocities in other media. The velocity of light in any medium of refractive index μ is—

$$\frac{\text{velocity in air}}{\mu}.$$

---

* "Annales de l'Observatoire de Paris," 1876. "Mémoires," tom. xiii. p. A_1.
† "Electricity," vol. ii. p. 235.

Professor Clerk Maxwell has proved mathematically that the velocity of electro-magnetic induction in any medium is—

$$\frac{\text{velocity in air}}{\sqrt{K}}$$

where K is the specific inductive capacity for electro-static induction as defined in vol. i. page 69.

Now, if the velocity of light is equal to that of electro-magnetic induction in all transparent insulators, we should have, .

$$\frac{\text{velocity of light in air}}{\mu} = \frac{\text{velocity of electro-magnetic induction in air}}{\sqrt{K}}$$

But we have shown that the velocities in air are equal, and hence, if the other velocities are equal, we must have—*

$$\mu = \sqrt{K}.$$

203. Gordon's experiments. The following table compares the values of μ and $\sqrt{K}$ for various dielectrics as determined by the present writer (see "Electricity," vol. i. page 118):—

Dielectric.	$\sqrt{K}$.	μ.
Double extra-dense flint glass	1·778	1·672
Extra-dense flint glass	1·747	1·620
Light flint glass	1·734	1·555
Hard crown glass	1·763	1·504
Paraffin	1·4119	1·4220

204. Gibson and Barclay's experiments. Messrs. Gibson and Barclay found for paraffin (see "Electricity," vol. i. page 86):—

$$\sqrt{K} = 1·405$$

---

* We must note that Professor Maxwell shows that among the values of μ we must select that which corresponds to waves of infinite wave length.

which does not differ much from the value of μ given in the preceding table.

205. Boltzmann's experiments. We can either compare $\sqrt{K}$ with μ or K with μ^2.

In the comparison given by Professor Boltzmann, the latter plan is adopted.

The following table, comparing the values of K and μ^2, is given by Professor Boltzmann in his paper quoted in "Electricity," vol. i. page 87.

Dielectric.	K.		μ^2.
	From condenser method.	From attraction method.	
Sulphur	3·84	3·90	4·06
Paraffin	2·32	{ 2·30 / 2·34 }	2·33
Resin	2·55	2·48	2·38

206. Crystalline Sulphur. In the paper quoted in "Electricity," vol. i. page 100, Professor Boltzmann gives the following comparison of K and μ^2 along the three axes g, m, k of crystalline sulphur.

Dielectric.		K.	μ^2.
Sulphur.	g	4·773	4·596
	m	3·970	3·886
	k	3·811	3·591

207. Schiller's experiments. In the paper quoted in "Electricity," vol. i. page 103, Schiller gives the following comparison :—

Dielectric.	K.		μ^2.
	By slow method.	By oscillation method.	
Paraffin, slow cooled, white	2·47	1·89 1·81	2·34 2·19*
Quickly cooled, nearly transparent	1·92	—	
Brown India Rubber	2·34	2·12	2·25

208. Silow's experiments. In the paper quoted in "Electricity," vol. i. page 104, Silow finds for oil of turpentine—

$$\sqrt{K} = 1\text{·}490. \quad \mu = 1\text{·}459.$$

209. Boltzmann's comparison for Gases. In the paper quoted in "Electricity," vol. i. page 123, Professor Boltzmann gives the following comparison for gases.

The refractive index and the specific inductive capacity of vacuum are taken as unity.

Dielectric—Gases at 0°C. and 760 mm.	$\sqrt{K}$.	μ.
Air	1·000295	1·000294
Carbonic acid	1·000473	1·000449
Hydrogen	1·000132	1·000138
Carbonic oxide	1·000345	1·000340
Nitrous gas (N.O.)	1·000497	1·000503
Olefiant gas	1·000656	1·000678
Marsh gas	1·000472	1·000443

* There is some confusion as to the arrangement of the numbers for paraffin in the table given in Schiller's paper.

210. General Conclusion. An examination of the foregoing tables shows us that in some cases the velocities of light and of electro-magnetic induction are very nearly equal, but that in other cases there is a very wide difference.

On the whole a sufficiently close agreement has been observed to give us fair hope that some day the discrepancies may be explained and eliminated; and meanwhile the **close** agreement of the velocities of light and **electro-magnetic** induction **in** air and in gases, and the numerous direct relations which exist between light and electricity leave us but little doubt that **they are very** closely related, and that **their** effects **are but** two forms **of** that **common** energy whose **nature is unknown,** but which certainly **underlies** all physical phenomena.

QUESTIONS ON CHAPTER XXVIII.

1. **State** Clerk Maxwell's electro-magnetic theory **of** light.

2. What arguments for **it do** we derive from experiments on the velocities of **light and** electro-magnetic **induction** in air?

3. What **arguments do we derive from** experiments on specific **inductive capacity?**

EXAMINATION PAPERS.

I.

1. State Ohm's law.
2. What current will 120 volts send through ·75 ohm?
3. Describe Grove's cell.
4. How many Watts are given out by a current of 18 ampères working at a pressure of 50 volts?
5. A coil of wire is wound round a bar of soft iron. What change occurs in the iron when a current passes round it?
6. Is whatever happens altered when the direction of the current is altered?
7. A long insulated wire is folded in the middle, and the two ends are held together. The centre, where the fold is, being secured to a soft iron bar, the double wire is wound round the bar so as to form a coil. The two free ends that project from the outside of the coil are separated, and a current sent through the wire. What happens to the iron bar?

[If the class have a difficulty in understanding this question, the method of winding may be illustrated by a piece of string and a ruler.]

8. What is the resistance of a circuit where 150 volts send 4000 ampères through it?
9. Describe the tangent galvanometer.
10. Give a rule for remembering which way a current deflects a compass needle.

II.

1. Describe the lecture model of Wheatstone's bridge.

2. In the figure—

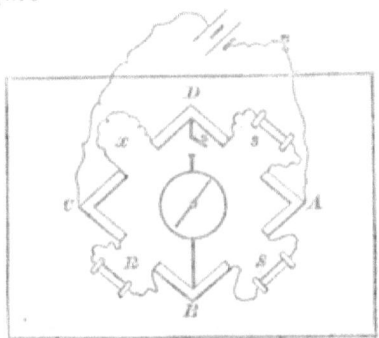

$S = s = 1000$, and the galvanometer remains at zero when $R = 701$, find x.

3. In the same figure—
$S = 1000$, $s = 10$, and the galvanometer remains at zero when $R = 702$, find x.

4. Describe Cardew's expansion volt-meter.
5. What is a volt?
6. What is the commercial electric unit?
7. How many horse-power are developed by 6 ampères flowing under a pressure of $124\frac{1}{3}$ volts.
8. Describe with a sketch the common electric bell.
9. Sketch a simple telegraph circuit.
10. Describe Daniell's cell.

III.

1. Describe the reflecting galvanometer.
2. What do we mean when we say a body is "diamagnetic"?
3. A little ball is hung by a long silk thread between the pointed poles of a powerful electro-magnet, but just out of the centre. What will happen on a current being sent round the coils of the magnet (*a*) when the ball is iron, (*b*) when it is bismuth?
4. What will happen if the ball is half iron and half bismuth? Give a reason for your answer.

5. Describe the Leyden jar.
6. Describe the Gramme ring.
7. How much horse-power is expended in 500 lamps, each taking ·8 ampère and 120 volts?
8. How much horse-power must the engine expend in producing this quantity of electricity, supposing the efficiency to be 82 per cent.?
9. State Lenz's law.
10. Sketch and describe Bell's first telephone.

IV.

1. The resistance of a circuit is 150 ohms, what pressure will be required to send $\frac{2}{3}$ of an ampère through it?
2. A dynamo machine produces 60 commercial units per hour, and the engine working it indicates 100 H.P., what is the efficiency?
[In this question the reciprocal of ·746 may be taken as $1\frac{1}{3}$.]
3. Describe the induction coil.
4. What is the difference in the spark when taken in air and in a partial vacuum?
5. Describe the phenomena of residual charge.
6. What arguments do we derive from it as to the way in which static induction is propagated?
7. State Faraday's discovery of the action of magnets on light.
8. How would you trace the lines of force of a magnet?
9. What is the difference between a conductor and an insulator?
10. What is an ampère?

V.

1. State the theory of artificial lighting.
2. Show how the same quantity of heat may produce different quantities of light under different circumstances. What are those circumstances?
3. What is electrolysis?

4. Describe the process of silver plating.

5. If electricity is being supplied at 6d. per unit, and if the lamps give 210 candles per H.P., and if gas giving 14 candles per 5 cubic feet per hour costs 2s. 10d. per 1000 cubic feet, what will be the cost of electricity for a quarter in a house where the quarter's gas-bill had previously been £12, assuming the same quantities of light to have been used in both cases?

[The reciprocal of ·746 may be taken as $1\frac{1}{3}$.]

6. What is radiant matter? how does it differ from ordinary gas?

7. State the theory of Wheatstone's bridge.

8. Describe some of Professor Bjerknes' experiments for showing how electrical attractions and repulsions can be imitated by vibrations in water.

9. If a closed ring of wire be moved rapidly in its own plane across the lines of the earth's magnetic force, what happens?

10. Describe the Hughes microphone.

VI.

1. Describe an experiment to show that radiant matter cannot turn a corner.

2. Describe and sketch the Crompton arc lamp.

3. Twenty H.P. is used in a group of 120 volt lamps. The main is a long one, and in order to keep 120 volts at the lamps the dynamo has to be worked at 150 volts. What H.P. is wasted in the main?

4. What is the resistance of the main?

5. Describe the process of making and laying a submarine cable.

6. Describe the condenser method of determining the electric pressure of a circuit.

7. What is magne-crystallic action?

8. What is a telephone exchange?

9. What is the difference between a "direct" and an "alternating" current?

10. If an electrically lighted district worked at 120 volts

uses 2000 ampères, what ampères will be required in a similar district using the same quantity of light but worked at 100 volts?

VII.

1. What is meant by "Dia-magnetic Polarity"?
2. Describe an experiment to prove its existence.
3. Describe Kerr's experiment for showing the action of static electricity on light.
4. A wire with a section of $\frac{1}{10}$ square inch is required to carry a certain current 200 yards with a fall of 15 volts, what section must the wire have to carry it 300 yards with the same fall?
5. What are the principal requisites of a good carbon for electric lighting?
6. State the laws of the attraction and repulsion between charged bodies, i.e. when will there be attraction, when repulsion, and how will the force be altered by alteration of distance?
7. Describe an experiment to show that tapping the glass of a Leyden jar hastens the return of the residual charge.
8. What do we learn from this experiment?
9. State Clerk Maxwell's electro-magnetic theory of light.
10. About what are the relative quantities of light per horse-power obtained from arc and incandescent lamps respectively?

VIII.

1. Describe Gordon's static induction balance.
2. What is a "dielectric"?
3. What is the difference between the natural and magnetic rotation of polarized light?
4. Describe an experiment to show that radiant matter produces heat when its motion is arrested.
5. Describe the Gordon dynamo.
6. If a ton of copper is required to convey 32 H.P. of electricity half a mile to 120 volt lamps, what weight will

be required to convey the same quantity of electricity to 90 volt lamps at the same distance with the same loss of H.P.

7. Describe an experiment to show that static induction can turn a corner.

8. What happens when one end of a magnet is plunged into the inside of a coil of wire? What when it is drawn out again?

9. Does the fact that electric signals are transmitted with great speed give us any information as to the speed of an electric current?

10. What is an astatic needle?

IX.

1. What are the properties of a magnet?
2. An electro-magnet wound in a right-handed spiral is placed upright on a table and a current sent round it in the direction of the hands of a watch. Will the N. pole be at the top or bottom?
3. Describe the process of manufacture of an incandescent lamp.
4. In a Wheatstone bridge test the galvanometer will not come to zero, but with the coils unplugged which would make x the resistance under test = 1215, it deflects 20 divisions to the right, and with those which would make x 1216, it deflects 5 divisions to the left. What is the value of x?
5. What are the three ordinary states of matter?
6. Give an illustration of the probable size of the molecules of water.
7. What horse-power is expended by 151 volts working through a resistance of 10 ohms?
8. What by the same volts working through a resistance of 1 ohm?
9. State the general theory of electric generators.
10. Describe Smee's cell.

s

X.

1. The hourly observations of a circuit kept at 150 volts give the following mean currents:—

Noon	to	2 p.m.	200 ampères.
2	,,	3 ,,	800 ,,
3	,,	5 ,,	3000 ,,
5	,,	6 ,,	2500 ,,
6	,,	7 ,,	1500 ,,
7	,,	11 ,,	1200 ,,
11	,,	12 mid. n.	1000 ,,
12	,,	6 a.m.	900 ,,
6	,,	8 ,,	1200 ,,
8	,,	10 ,,	150 ,,
10	,,	12 noon	170 ,,

How many units have been supplied in the 24 hours?

2. Describe the Siemens electro-dynamometer.

3. Show how the coils are grouped in a bridge resistance box, so that with 16 coils any resistance from 10,000 ohms to 1 ohm can be obtained.

4. Describe the Gower-Bell telephone.

5. Describe Faraday's voltameter.

6. Describe Gordon's experiments for examining the effect of variations of pressure on the lengths of discharges in air.

7. Describe an experiment to show that radiant matter exerts mechanical force where it strikes.

8. Describe Bjerknes' method of tracing lines of force in water.

9. What is the effect of connecting the coatings of a Leyden jar to the terminals of an induction coil?

10. Give a mechanical illustration of what occurs.

INDEX.

A

ALTERNATING current **machines**, 133.
———— and direct currents, 133.
Amalgamated zinc, 35.
Ampère, 21.
Ampère's theory **of magnetism**, 111.
Arc lamps, 121.
Astatic needles, 53.
Ayrton and Perry on specific inductive capacity of gases, 225.

B

BATTERIES, 34.
———— of several cells, 41.
Bells, electric, 102.
Bell's telephone, 96.
Bjerknes' experiments on mechanical analogies of electrical phenomena, 227.
Board of Trade unit, the, 31.
Boltzmann on specific inductive capacity of gases, 225.
Burgin machine, 137.

C

CAPACITY, specific inductive, 215.
Carbon, use of, in electric lighting, 112.
Carbons for arc lamps, 128.

Cardew's voltmeter, 69.
Cell, voltaic theory of, 34.
Charges of electricity, 201.
Clausius' **theory** of electrolysis, 145.
Coil, induction, 162.
Commercial Unit, the, 31.
Condenser for induction coil, primary, 165.
————————, secondary, 169.
———— method **of** measuring pressure, 63, 64.
Conductors and insulators, **1.**
Constant **batteries,** 36.
Coulomb, 23.
Crompton's arc lamp, **125.**
Crookes **on** radiant matter, 182.
Current, effects of, **2.**
————, unit of, 21.
————, horse-power and resistance, 25.
———————— pressure, 26.
Currents **and** magnets, **8.**
———— mechanical forces, 17.
————, conversion of into heat, 108.
————, measurement of, 50.

D

DANIELL'S cell, 28.
Diamagnetic polarity, 155.
Diamagnetism, 153.
Direct and alternating currents, 133.

Direct current machines, 137.
────── method of measuring pressure, 67.
Discharge in rarefied air, 171.
────── of the induction coil, and discharge generally, 168.
Dynamo machines, 48, 133.

E

EARTH-PLATES, 87.
Effects of the electric current, 2.
Efficiency of incandescent lamps, 119.
Electro-dynamometer, Siemens' 58.
Electrolysis, 143.
Electro-magnetic induction, 43.
────── theory of light, 245.
Electro-magnets, 10, 150.
Electro-motive force, see Pressure.
Electro-plating, 148.
Electro-static action on light, 243.
────── induction, 201.
Energy and horse-power, 24.
──────, unit of, 32.
Examination papers, 252.

F

FARADAY on action of magnets on light, 237.
────── specific inductive capacity, 216.
Field magnets, magnetization of, 140.
Foot-pounds, 18.
Formulæ, summary of, 30.

G

GALVANOMETER shunts, 82.
Galvanometers, 50.
Gas and electric units, 32.

Gases, specific inductive capacity of, 225.
Generators of electricity, 45.
Gordon on specific inductive capacity, 219.
Gordon's dynamo, 133.
Gower-Bell telephone, 98.
Gramme ring, 137.
Grove's cell, 37.

H

HEAT and work, units of, 24.
────── produced by electric current, 110.
Horse-power, 18, 24.
──────, current, and pressure, 26.
──────, resistance, and current, 25.
────── pressure, 27.

I

INCANDESCENT lamps, 115.
Induced charges of electricity, 203.
Induction balance, the static, 219.
────── by variation of current, 47.
──────, electro-magnetic, 43.
──────, electro-static, 201.
────── precedes discharge, 211.
────── coil, 162.
────── and magneto machine, 171.
Inductive capacity, specific, 215.
Insulators and conductors, 1.
Iron core, 47.

K

KERR on action of electricity on light, 243.

L

Lamps, electric, theory of, 111.
Land telegraphy, 86.
Latimer Clark's standard cell, 39.
Law of force between charged bodies, 202.
Laws of electrolysis, 144.
Lenz's law, 47.
Leyden jar, the, 205.
Light acted on by **magnet**, 237.
———, electro-static **action on**, 243.
———, Maxwell's electro-magnetic theory of, 245.
———, polarized, 236.
Lighting, electric, 105.
Line of force, Maxwell's, 246.
Lines of force, 12, 212.
——————, Bjerknes', 233.

M

Magne-crystallic action, 157.
Magnets, 7.
————, action of, on **light, 237.**
———— and currents, 8.
Matter, on the nature of, 179.
Maxwell's electro-magnetic theory of light, 245.
Mechanical forces and currents, 17.
Metchim's electrometer, 68.
Microphone, 97.
Molecules, size of, 180.
Morse code, 89.
———— instrument, 88.

O

Ohm, the, 22.
Ohm's law, 4, 23.
Overhead wires, 86.

P

Phases, table of, 230, 231.
Polarized light, 236.
Pressure, current and horse-power, 26.
————, measurement of, 62.
————, resistance and horse-power, 27.
————, unit of, 22.
————, work, and quantity, 28.

Q

Quantity, unit of, 23.
————, work, and pressure, 28.

R

Radiant matter, 181.
Reflecting galvanometers, 55.
Residual charge, 206.
Resistance box, 77.
———————— coils, 77.
————————, horse-power, and current, 25.
———————————————— - pressure, 27.
————————, measurement of, 71.
————————, specific, 80.
————————, unit of, 22.
Rotation of polarized light, **magnetic**, 237.
————————————————, natural, 237.

S

Secondary condenser, 169.
Sensitive galvanometers, 53.
Serrin's arc lamp, 122.
Shunts, 82.
Siemens' armature, 137, 140.
———————— electro-dynamometer, 58.

Sign for battery, 42.
Silver-plating, 148.
Single-needle instrument, 87.
Size of molecules, 180.
Small current method of measuring pressure, 69.
Spark-length, experiments on, 171.
Specific inductive capacity, 215.
——— resistances, table of, 81.
Speed of electric currents, 5.
Spottiswoode's coil, 166.
Strain, how propagated, 210.
——— of insulator by induction, 203.
Striæ, 176.
Stroh's experiments on analogies between air vibrations and electrical phenomena, 234.
Submarine telegraphy, 90.
Swan's incandescent lamps, 115.

Torsion balance for electro-magnets, 152.
Two-fluid cells, 37.

U

UNDERGROUND wires, 87.
Unit, the Commercial, 31.
Units, electrical, and their relation to each other, 21.

V

VACUUM tubes, 176.
Velocities of light and electro-magnetic induction compared, 246.
Volt, the, 22.
Voltameter, the, 147.
Voltmeters, 69.

T

TANGENT galvanometer, 50.
Telegraphy, 11, 86.
Telephone, Bell's, 96.
———, Gower-Bell, 98.
——— exchanges, 100.
Theory of electric generators, 45.
——— ——— lamps, 111.
Thomson's galvanometer, 55.

W

WATT, the, 32.
Wheatstone's bridge, 71.
Work, quantity, and pressure, 28.

Y

YOKES and magnetic circuits, 14.

THE END.

LONDON:
PRINTED BY GILBERT AND RIVINGTON, LIMITED,
ST. JOHN'S SQUARE.

A Catalogue of American and Foreign Books Published or Imported by MESSRS. SAMPSON LOW & CO. *can be had on application.*

Crown Buildings, 188, Fleet Street, London,
October, 1885.

A Selection from the List of Books

PUBLISHED BY

SAMPSON LOW, MARSTON, SEARLE, & RIVINGTON.

ALPHABETICAL LIST.

ABOUT Some Fellows. By an ETON BOY, Author of "A Day of my Life." Cloth limp, square 16mo, 2s. 6d.

Adams (C. K.) Manual of Historical Literature. Cr. 8vo, 12s. 6d.

Alcott (Louisa M.) Jack and Jill. 16mo, 5s.
—— —— *Old-Fashioned Thanksgiving Day.* 3s. 6d.
—— —— *Proverb Stories.* 16mo, 3s. 6d.
—— —— *Spinning-Wheel Stories.* 16mo, 5s.
—— —— See also "Rose Library."

Alden (W. L.) Adventures of Jimmy Brown, written by himself. Illustrated. Small crown 8vo, cloth, 2s. 6d.

Aldrich (T. B.) Friar Jerome's Beautiful Book, &c. Very choicely printed on hand-made paper, parchment cover, 3s. 6d.
—— —— *Poetical Works.* Édition de Luxe. 8vo, 21s.

Alford (Lady Marian) Needlework as Art. With over 100 Woodcuts, Photogravures, &c. Royal 8vo, 42s.; large paper, 84s.

Amateur Angler's Days in Dove Dale: Three Weeks' Holiday in July and August, 1884. By E. M. Printed by Whittingham, at the Chiswick Press. Cloth gilt, 1s. 6d.; fancy boards, 1s.

American Men of Letters. Thoreau, Irving, Webster. 2s. 6d. each.

Anderson (W.) *Pictorial Arts of Japan.* With 80 full-page and other Plates, 16 of them in Colours. Large imp. 4to, gilt binding, gilt edges, 8l. 8s.; or in four parts, 2l. 2s. each.

Angler's Strange Experiences (An). By COTSWOLD ISYS. With numerous Illustrations, 4to, 5s. New Edition, 3s. 6d.

Angling. See Amateur, "British Fisheries Directory," "Cutcliffe," "Martin," "Stevens," "Theakston," "Walton," and "Wells."

Arnold (Edwin) Birthday Book. 4s. 6d.

A

Art Education. See "Biographies of Great Artists," "Illustrated Text Books," "Mollett's Dictionary."

Artists at Home. Photographed by J. P. MAYALL, and reproduced in Facsimile. Letterpress by F. G. STEPHENS. Imp. folio, 42s.

Audsley (G. A.) Ornamental Arts of Japan. 90 Plates, 74 in Colours and Gold, with General and Descriptive Text. 2 vols., folio, £15 15s. On the issue of Part III. the price will be further advanced.

────── *The Art of Chromo-Lithography.* Coloured Plates and Text. Folio, 63s.

Auerbach (B.) Brigitta. Illustrated. 2s.

────── *On the Heights.* 3 vols., 6s.

────── *Spinoza.* Translated. 2 vols., 18mo, 4s.

BALDWIN (*J.*) *Story of Siegfried.* 6s.

────── *Story of Roland.* Crown 8vo, 6s.

Ballin (Ada S., Lecturer to the National Health Society) Science of Dress in Theory and Practice. Illustrated, 6s.

Barlow (Alfred) Weaving by Hand and by Power. With several hundred **Illustrations**. Third Edition, royal 8vo, 1l. 5s.

Barlow (*William*) *New Theories of Matter and Force.* 2 vols., 8vo,

THE BAYARD SERIES.
Edited by the late J. HAIN FRISWELL.

Comprising Pleasure Books of Literature produced in the Choicest Style as Companionable Volumes at Home and Abroad.

"We can hardly imagine better books for boys to read or for men to ponder over."—*Times.*

Price 2s. 6d. each Volume, complete in itself, flexible cloth extra, gilt edges, with silk Headbands and Registers.

The Story of the Chevalier Bayard. By M. De Berville.

De Joinville's St. Louis, King of France.

The Essays of Abraham Cowley, including all his Prose Works.

Abdallah; or, The Four Leaves. By Edouard Laboullaye.

Table-Talk and Opinions of Napoleon Buonaparte.

Vathek: An Oriental Romance. By William Beckford.

Words of Wellington: Maxims and Opinions of the Great Duke.

Dr. Johnson's Rasselas, Prince of Abyssinia. With Notes.

Hazlitt's Round Table. With Biographical Introduction.

The Religio Medici, Hydriotaphia, and the Letter to a Friend. By Sir Thomas Browne, Knt.

Ballad Poetry of the Affections. By Robert Buchanan.

Coleridge's Christabel, and other Imaginative Poems. With Preface by Algernon C. Swinburne.

Lord Chesterfield's Letters, Sentences, and Maxims. With Introduction by the Editor, and

Bayard *Series (continued)* :—

Essay on Chesterfield by M. de Ste.-Beuve, of the French Academy.

The King and the Commons. A Selection of Cavalier and Puritan Songs. Edited by Professor Morley.

Essays in Mosaic. By Thos. Ballantyne.

My Uncle **Toby**; his Story and his Friends. Edited by P. Fitzgerald.

Reflections; or, Moral Sentences and Maxims of the Duke de la Rochefoucauld.

Socrates: Memoirs for English Readers from Xenophon's Memorabilia. By Edw. Levien.

Prince Albert's Golden Precepts.

A Case containing 12 Volumes, price 31s. 6d.; or the Case separately, price 3s. 6d.

Behnke *and Browne*. ***Child's Voice***. Small 8vo, 3s. 6d.

Bickersteth *(Bishop E. H.)* **The Clergyman in his Home.** Small post 8vo, 1s.

———— *Evangelical Churchmanship and Evangelical Eclecticism.* 8vo, 1s.

———— *From Year to Year: Original Poetical Pieces.* Small post 8vo, 3s. 6d.; roan, 6s. and 5s.; calf or morocco, 10s. 6d.

———— *Hymnal Companion to the Book of Common Prayer.* May be had in various styles and bindings from 1d. to 31s. 6d. *Price List and Prospectus will be forwarded on application.*

———— **The** *Master's Home-Call; or, Brief Memorials of Alice* Frances Bickersteth. 20th Thousand. 32mo, cloth gilt, 1s.

———— *The Master's* **Will.** **A Funeral Sermon** preached on the Death of Mrs. S. Gurney Buxton. Sewn, 6d.; cloth gilt, 1s.

———— *The* **Reef, and other Parables.** Crown 8vo, 2s. 6d.

———— *The Shadow of* **the Rock.** **A Selection of** Religious Poetry. 18mo, cloth extra, 2s. 6d.

———— *The Shadowed* **Home and the Light Beyond.** New Edition, crown 8vo, **cloth extra, 5s.**

Biographies of the Great **Artists** *(Illustrated).* **Crown** 8vo, emblematical binding, 3s. 6d. per volume, except where the price is given.

Claude Lorrain.*
Correggio, by M. E. Heaton, 2s. 6d.
Della Robbia and Cellini, 2s. 6d.
Albrecht **Dürer**, by R. F. Heath.
Figure Painters of Holland.
Fra Angelico, Masaccio, and Botticelli.
Fra Bartolommeo, Albertinelli, and Andrea del Sarto.
Gainsborough and Constable.
Ghiberti and Donatello, 2s. 6d.
Giotto, by Harry Quilter.
Hans Holbein, by Joseph Cundall.
Hogarth, by Austin Dobson.
Landseer, by F. G. Stevens.
Lawrence and Romney, by Lord Ronald Gower, 2s. 6d.

* *Not yet published.*

Biographies of the Great Artists (continued) :—

Leonardo da Vinci.
Little Masters of Germany, by W. B. Scott.
Mantegna and Francia.
Meissonier, by J. W. Mollett, 2s. 6d.
Michelangelo Buonarotti, by Clément.
Murillo, by Ellen E. Minor, 2s. 6d.
Overbeck, by J. B. Atkinson.
Raphael, by N. D'Anvers.
Rembrandt, by J. W. Mollett.
Reynolds, by F. S. Pulling.
Rubens, by C. W. Kett.
Tintoretto, by W. R. Osler.
Titian, by R. F. Heath.
Turner, by Cosmo Monkhouse.
Vandyck and Hals, by P. R. Head
Velasquez, by E. Stowe.
Vernet and Delaroche, by J. Rees.
Watteau, by J. W. Mollett, 2s. 6d.
Wilkie, by J. W. Mollett.

Bird (F. J.) American Practical Dyer's Companion. 8vo, 42s.

Bird (H. E.) Chess Practice. 8vo, 2s. 6d.

Black (Wm.) Novels. See "Low's Standard Library."

Blackburn (Charles F.) Hints on Catalogue Titles and Index Entries, with a Vocabulary of Terms and Abbreviations, chiefly from Foreign Catalogues. Royal 8vo, 14s.

Blackburn (Henry) Breton Folk. With 171 Illust. by RANDOLPH CALDECOTT. Imperial 8vo, gilt edges, 21s.; plainer binding, 10s. 6d.

—— *Pyrenees (The).* With 100 Illustrations by GUSTAVE DORÉ, corrected to 1881. Crown 8vo, 7s. 6d.

Blackmore (R. D.) Lorna Doone. *Édition de luxe.* Crown 4to, very numerous Illustrations, cloth, gilt edges, 31s. 6d.; parchment, uncut, top gilt, 35s. Cheap Edition, small post 8vo, 6s.

—— *Novels.* See "Low's Standard Library."

Blaikie (William) How to get Strong and how to Stay so. Rational, Physical, Gymnastic, &c., Exercises. Illust., sm. post 8vo, 5s.

—— *Sound Bodies for our Boys and Girls.* 16mo, 2s. 6d.

Bonwick (Jos.) British Colonies and their Resources. 1 vol., cloth, 5s. Sewn—I. Asia, 1s.; II. Africa, 1s.; III. America, 1s.; IV. Australasia, 1s.

Bosanquet (Rev. C.) Blossoms from the King's Garden: Sermons for Children. 2nd Edition, small post 8vo, cloth extra, 6s.

Boussenard (L.) Crusoes of Guiana. Illustrated. 5s.

—— *Gold-seekers, a Sequel.* Illustrated. 16mo, 5s.

Boy's Froissart. King Arthur. Mabinogion. Percy. See LANIER.

Bradshaw (J.) New Zealand as it is. 8vo, 12s. 6d.

Brassey (Lady) Tahiti. With 31 Autotype Illustrations after Photos. by Colonel STUART-WORTLEY. Fcap. 4to, 21s.

Bright (John) Public Letters. Crown 8vo, 7s. 6d.

Brisse (Baron) Ménus (366). A *ménu*, in French and English, for every Day in the Year. Translated by Mrs. MATTHEW CLARKE. 2nd Edition. Crown 8vo, 5s.

British Fisheries Directory, 1883-84. Small 8vo, **2s.** 6d.

Brittany. See BLACKBURN.

Brown. Life and Letters of John Brown, Liberator of Kansas, and Martyr of Virginia. By F. B. SANBORN. Illustrated. 8vo, 12s. 6d.

Browne (G. Lennox) Voice Use and Stimulants. Sm. 8vo, 3s. 6d.

―――― and Behnke (*Emil*) *Voice, Song, and Speech.* Illustrated, 3rd Edition, medium 8vo, 15s.

Bryant *(W. C.) and Gay (S. H.) History of the United States.* 4 vols., royal 8vo, profusely Illustrated, 60s.

Bryce (Rev. Professor) Manitoba. With Illustrations **and Maps.** Crown 8vo, **7s. 6d.**

Bunyan's Pilgrim's Progress. With 138 original **Woodcuts.** Small **post** 8vo, **cloth** gilt, **3s. 6d.;** gilt edges, 4s.

Burnaby (Capt.) On Horseback through Asia Minor. **2** vols., 8vo, 38s. Cheaper Edition, **1** vol., crown 8vo, 10s. 6d.

Burnaby (Mrs. F.) High Alps in Winter; or, Mountaineering in Search of Health. By Mrs. FRED BURNABY. With Portrait of the Authoress, Map, and other Illustrations. Handsome cloth, 14s.

Butler (W. F.) The Great Lone Land; an Account of **the** Red River Expedition, 1869-70. New Edition, cr. 8vo, cloth extra, 7s. 6d.

―――― *Invasion of England, told twenty years after, by* **an** *Old* Soldier. Crown 8vo, 2s. 6d.

―――― *Red Cloud; or,* **the** *Solitary Sioux.* Imperial 16mo, **numerous** illustrations, gilt edges, 5s.

―――― *The Wild North Land; the Story of a Winter Journey* with Dogs across Northern North America. 8vo, 18s. Cr. 8vo, **7s.** 6d.

Buxton (H. ***J. W.****) Painting, English and American.* Crown 8vo, 5s.

*C*ADOGAN *(Lady A.)* ***Illustrated Games*** *of Patience.* Twenty-four Diagrams in **Colours, with Text.** Fcap. 4to, **12s.** 6d.

California. See "**Nordhoff."**

Cambridge Staircase (A). By the Author **of "A Day of my** Life at Eton." Small crown 8vo, cloth, 2s. 6d.

Cambridge Trifles; from an Undergraduate Pen. By the Author of "A Day of my Life at Eton," &c. 16mo, cloth extra, 2s. 6d.

Carleton (Will) Farm Ballads, Farm Festivals, and Farm Legends. 1 vol., small post 8vo, 3s. 6d.

—————— *City Ballads.* With Illustrations. 12s. 6d.

—————— See also "Rose Library."

Carnegie (A.) American Four-in-Hand in Britain. Small 4to, Illustrated, 10s. 6d. Popular Edition, 1s.

—————— *Round the World.* 8vo, 10s. 6d.

Chairman's Handbook (The). By R. F. D. PALGRAVE, Clerk of the Table of the House of Commons. 5th Edition, 2s.

Changed Cross (The), and other Religious Poems. 16mo, 2s. 6d.

Charities of London. See Low's.

Chattock (R. S.) Practical Notes on Etching. Sec. Ed., 8vo, 7s. 6d.

Chess. See BIRD (H. E.).

Children's **Praises.** *Hymns for Sunday-Schools and Services.* Compiled by LOUISA H. H. TRISTRAM. 4d.

Choice *Editions of Choice Books.* 2s. 6d. each. Illustrated by C. W. COPE, R.A., T. CRESWICK, R.A., E. DUNCAN, BIRKET FOSTER, J. C. HORSLEY, A.R.A., G. HICKS, R. REDGRAVE, R.A., C. STONEHOUSE, F. TAYLER, G. THOMAS, H. J. TOWNSHEND, E. H. WEHNERT, HARRISON WEIR, &c.

Bloomfield's Farmer's Boy.	Milton's L'Allegro.
Campbell's Pleasures of Hope.	Poetry of Nature. Harrison Weir.
Coleridge's Ancient Mariner.	Rogers' (Sam.) Pleasures of Memory.
Goldsmith's Deserted Village.	Shakespeare's Songs and Sonnets.
Goldsmith's Vicar of Wakefield.	Tennyson's May Queen.
Gray's **Elegy in a** Churchyard.	Elizabethan Poets.
Keat's **Eve of St.** Agnes.	Wordsworth's Pastoral Poems.

"Such works are a glorious beatification for a poet."—*Athenæum.*

Christ in Song. By PHILIP SCHAFF. New Ed., gilt edges, 6s.

Chromo-Lithography. See "Audsley."

Collingwood (Harry) **Under the Meteor** *Flag.* The Log of a Midshipman. Illustrated, small post 8vo, gilt, 6s.; plainer, 5s.

—————— *The Voyage of the "Aurora."* Illustrated, small post 8vo, gilt, 6s.; plainer, 5s.

Colvile (H. E.) Accursed **Land***: Water Way of* **Edom.** 10s. 6d.

Composers. See "Great Musicians."

Confessions of a Frivolous Girl. Cr. 8vo, 6s. Paper boards, 1s.

Cook (Dutton) Book of the **Play.** New Edition. 1 vol., 3s. 6d.

—— *On the Stage: Studies of Theatrical History and the Actor's Art.* 2 vols., 8vo, cloth, **24s.**

Costume. **See Smith** (J. Moyr).

Cowen (Jos., **M.P.**) *Life and Speeches.* By Major Jones. 8vo, 14s.

Curtis (C. B.) Velazquez and Murillo. With Etchings, &c. Royal 8vo, 31s. 6d.; large paper, 63s.

Custer (E. B.) Boots and Saddles. Life in **Dakota** *with General Custer.* Crown 8vo, 8s. 6d.

Cutcliffe (H. C.) Trout Fishing in Rapid Streams. **Cr. 8vo, 3s. 6d.**

*D*ANVERS **(N.)** *An Elementary History of Art.* Crown 8vo, 10s. 6d.

—— *Elementary History of Music.* Crown 8vo, 2s. 6d.

—— *Handbooks of Elementary Art—Architecture; Sculpture; Old Masters; Modern Painting.* **Crown** 8vo, 3s. 6d. each.

Davis (C. T.) **Manufacture** *of Bricks,* **Tiles,** *Terra-Cotta, &c.* Illustrated. **8vo, 25s.**

—— *Manufacture of* **Leather.** With many Illustrations. 52s. 6d.

Dawidowsky (F.) **Glue, Gelatine,** *Isinglass, Cements, &c.* 8vo, 12s. 6d.

Day of My Life (A); or, Every-Day Experiences at Eton. By an Eton Boy. 16mo, cloth extra, 2s. 6d.

Day's *Collacon: an Encyclopædia of Prose Quotations.* Imperial 8vo, cloth, 31s. 6d.

Decoration. Vols. II. to IX. New Series, folio, 7s. 6d. each.

Dogs *in Disease: their Management and Treatment.* By Ashmont. Crown 8vo, **7s.** 6d.

Donnelly (Ignatius) **Atlantis; or,** *the Antediluvian World.* 7th Edition, crown **8vo,** 12s. **6d.**

—— *Ragnarok: The Age of* **Fire** *and* **Gravel.** Illustrated, **Crown** 8vo, 12s. 6d.

Doré (Gustave) Life and Reminiscences. By BLANCHE ROOSE-
VELT. With numerous Illustrations from the Artist's previously un-
published Drawings. Medium 8vo, 24*s.*

Dougall (James Dalziel) Shooting: its Appliances, Practice,
and Purpose. New Edition, revised with additions. Crown 8vo, 7*s.* 6*d.*
"The book is admirable in every way. .... We wish it every success."—*Globe.*
"A very complete treatise. .... Likely to take high rank as an authority on shooting."—*Daily News.*

Drama. See COOK (DUTTON).

Dyeing. See BIRD (F. J.).

EDUCATIONAL Works published in Great Britain. A
Classified Catalogue. Second Edition, 8vo, cloth extra, 5*s.*

Egypt. See "De Leon," "Foreign Countries."

Eight Months on the Gran Ciacco of the Argentine Republic.
8vo, 12*s.* 6*d.*

Electricity. See GORDON.

Elliot (Adm. Sir G.) Future Naval Battles, and how to Fight
them. Numerous Illustrations. Royal 8vo, 14*s.*

Emerson (R. W.) Life. By G. W. COOKE. Crown 8vo, 8*s.* 6*d.*

English Catalogue of Books. Vol. III., 1872—1880. Royal
8vo, half-morocco, 42*s.* See also "Index."

English Etchings. A Periodical published Monthly.

English Philosophers. Edited by E. B. IVAN MÜLLER, M.A.
A series intended to give a concise view of the works and lives of English
thinkers. Crown 8vo volumes of 180 or 200 pp., price 3*s.* 6*d.* each.

Francis Bacon, by Thomas Fowler.
Hamilton, by W. H. S. Monck.
Hartley and James Mill, by G. S. Bower.
*John Stuart Mill, by Miss Helen Taylor.
Shaftesbury and Hutcheson, by Professor Fowler.
Adam Smith, by J. A. Farrer.

* *Not yet published.*

Esmarch (Dr. Friedrich) Treatment of the Wounded in War.
Numerous Coloured Plates and Illust., 8vo, strongly bound, 1*l.* 8*s.*

Etching. See CHATTOCK, and ENGLISH ETCHINGS.

Etchings (Modern) of Celebrated Paintings. 4to, 31*s.* 6*d.*

FARM *Ballads, Festivals, and Legends.* See "Rose Library."

Fauriel (Claude) **Last** *Days of the Consulate.* Cr. 8vo, **10s. 6d.**

Fawcett (Edgar) **A** *Gentleman of Leisure.* 1s.

Feilden (H. St. C.) Some Public **Schools,** *their Cost* **and** Scholarships. Crown 8vo, 2s. 6d.

Fenn (G. Manville) Off to the Wilds: **A** *Story for Boys.* Profusely Illustrated. Crown 8vo, 7s. 6d. ; also 5s.

——— *The Silver Cañon:* **a Tale of the** *Western Plains.* Illustrated, small post 8vo, gilt, **6s.**; plainer, 5s.

Fennell (Greville) Book of the Roach. New Edition, 12mo, **2s.**

Ferns. **See HEATH.**

Fields (J. T.) Yesterdays with Authors. New Ed., 8vo, 10s. 6d.

Fleming (Sandford) England and Canada: **a** *Summer Tour.* Crown 8vo, 6s.

Florence. **See "Yriarte."**

Folkard (R., Jun.) **Plant** *Lore, Legends,* **and** *Lyrics.* Illustrated, 8vo, **16s.**

Forbes *(H. O.) Naturalist's Wanderings* **in the Eastern** *Archi*pelago. Illustrated, 8vo, **21s.**

Foreign **Countries** *and British Colonies.* A series of Descriptive Handbooks. Crown 8vo, 3s. 6d. each.

Australia, by J. F. Vesey Fitzgerald.
Austria, by D. **Kay,** F.R.G.S.
*Canada, by W. **Fraser** Rae.
Denmark and Iceland, by E. C. Otté.
Egypt, by S. Lane Poole, B.A.
France, by Miss M. Roberts.
Germany, by S. Baring-Gould.
Greece, by L. Sergeant, B.A.
*Holland, by **R.** L. Poole.
Japan, by S. Mossman.
*New Zealand.
*Persia, by Major-Gen. **Sir F. Gold**smid.
Peru, by Clements R. Markham, C.B.
Russia, by W. R. Morfill, M.A.
Spain, by Rev. Wentworth Webster.
Sweden and Norway, by F. H. Woods.
*Switzerland, by W. A. P. Coolidge, M.A.
*Turkey-in-Asia, **by J. C. McCoan,** M.P.
West Indies, **by C. H. Eden,** F.R.G.S.

* *Not ready yet.*

Frampton (Mary) **Journal, Letters,** *and Anecdotes,* 1799— 1846. 8vo, 14s.

Franc (Maud Jeanne). The following form one Series, small post 8vo, in uniform cloth bindings, with gilt edges :—

Emily's Choice. 5s.	Vermont Vale. 5s.
Hall's Vineyard. 4s.	Minnie's Mission. 4s.
John's Wife: A Story of Life in South Australia. 4s.	Little Mercy. 4s.
	Beatrice Melton's Discipline. 4s.
Marian; or, The Light of Some One's Home. 5s.	No Longer a Child. 4s.
	Golden Gifts. 4s.
Silken Cords and Iron Fetters. 4s.	Two Sides to Every Question. 4s.
Into the Light. 4s.	Master of Ralston, 4s.

Francis (Frances) Elric and **Ethel: a Fairy** *Tale.* Illustrated. Crown 8vo, 3s. 6d.

French. See "Julien."

Froissart. See "Lanier."

GALE (F.; the Old Buffer) Modern **English** *Sports: their* Use and Abuse. Crown 8vo, 6s.; a few large paper copies, 10s. 6d.

Garth **(Philip)** *Ballads and Poems from the* **Pacific.** Small post 8vo, 6s.

Gentle Life (Queen Edition). 2 vols. in 1, small 4to, 6s.

THE GENTLE LIFE SERIES.

Price 6s. each; or in calf extra, price 10s. 6d.; Smaller Edition, cloth extra, 2s. 6d., except where price is named.

The Gentle Life. Essays in aid of the Formation of Character of Gentlemen and Gentlewomen.

About in the World. Essays by Author of "The Gentle Life."

Like unto Christ. A New Translation of Thomas à Kempis' "De Imitatione Christi."

Familiar Words. An Index Verborum, or Quotation Handbook. 6s.

Essays by Montaigne. Edited and Annotated by the Author of "The Gentle Life."

The Gentle Life. 2nd Series.

The Silent Hour: Essays, Original and **Selected.** By the Author of "The Gentle Life."

Half-Length **Portraits.** Short Studies of Notable Persons. By J. HAIN FRISWELL.

Essays on English Writers, **for** the Self-improvement **of** Students in English Literature.

Other People's Windows. **By** J. HAIN FRISWELL. **6s.**

A Man's Thoughts. **By** J. HAIN FRISWELL.

The Countess of **Pembroke's** *Arcadia.* By Sir PHILIP SIDNEY. New Editior, **6s.**

George Eliot: a Critical Study of her Life. **By** G. W. COOKE. Crown 8vo, 10s. 6d.

Germany. By S. BARING-GOULD. Crown 8vo, 3s. **6d.**

Gilder (W. H.) Ice-Pack and Tundra. An Account of the Search for the "Jeannette." 8vo, 18s.

———— *Schwatka's Search.* Sledging in quest of the Franklin Records. Illustrated, 8vo, 12s. 6d.

Gilpin's Forest Scenery. **Edited by F. G. HEATH.** Post 8vo, 7s. 6d.

Gisborne **(W.)** *New Zealand* **Rulers and** *Statesmen.* With Portraits. Crown 8vo,

Gordon *(General)* **Private** *Diary in China.* Edited **by** S. MOSSMAN. Crown **8vo, 7s.** 6d.

Gordon (J. E. H., B.A. Cantab.) Four Lectures on Electric Induction **at the Royal** Institution, 1878-9. Illust., square 16mo, 3s.

———— **Electric Lighting.** Illustrated, **8vo,** 18s.

———— *Physical* **Treatise on** *Electricity and Magnetism.* **2nd** Edition, enlarged, with coloured, full-page, &c., Illust. 2 vols., 8vo, 42s.

———— *Electricity for Schools.* Illustrated. Crown 8vo, 5s.

Gouffé (Jules) Royal Cookery Book. Translated and adapted for English use by ALPHONSE GOUFFÉ, Head Pastrycook to the Queen. New Edition, with plates in colours, Woodcuts, &c., 8vo, gilt edges, **42s.**

———— **Domestic Edition,** half-bound, 10s. **6d.**

Grant (General, U.S.) Personal Memoirs. With numerous Illustrations, Maps, &c. **2 vols.,** 8vo, 28s.

Great *Artists.* See "Biographies."

Great Musicians. Edited by F. HUEFFER. A Series of Biographies, crown 8vo, 3s. each :—

Bach.	Handel.	Purcell.
*Beethoven.	Haydn.	Rossini.
*Berlioz.	*Marcello.	Schubert.
English Church Composers. By BARETT.	Mendelssohn.	Schumann.
	Mozart.	Richard Wagner.
*Gluck.	*Palestrina.	Weber.

* *In preparation.*

Groves (J. Percy) Charmouth Grange: a Tale of the Seventeenth Century. Illustrated, small post 8vo, gilt, 6s.; plainer, 5s.

Guizot's History of France. Translated by ROBERT BLACK. Super-royal 8vo, very numerous Full-page and other Illustrations. In 8 vols., cloth extra, gilt, each 24s. This work is re-issued in cheaper binding, 8 vols., at 10s. 6d. each.

"It supplies a want which has long been felt, **and ought to be in the hands of all students of history.**"—*Times.*

—————————— *Masson's School Edition.* Abridged from the Translation by Robert Black, with Chronological Index, Historical and Genealogical Tables, &c. By Professor GUSTAVE MASSON, B.A. With **24** full-page Portraits, and other Illustrations. 1 vol., 8vo, 600 pp., 10s. 6d.

Guizot's History of England. In 3 vols. of about 500 pp. each, containing 60 to 70 full-page and other Illustrations, cloth extra, gilt, 24s. each; re-issue in cheaper binding, 10s. 6d. each.

"**For** luxury of typography, plainness of print, and beauty of illustration, these **volumes,** of which but one has as yet appeared in English, will hold their own against any production of an age so luxurious as our own in everything, typography not excepted."—*Times.*

Guyon (Mde.) Life. By UPHAM. 6th Edition, crown 8vo, 6s.

HALFORD (F. M.) Floating Flies, and how to Dress them. Coloured plates. 8vo, 15s.; large paper, 30s.

Hall (W. W.) How to Live Long; or, 1408 *Heatth Maxims,* Physical, Mental, and Moral. 2nd Edition, small post 8vo, 2s.

Hamilton (E.) Recollections of Fly-fishing for Salmon, Trout, and Grayling. With their Habits, Haunts, and History. Illustrated, small post 8vo, 6s.; large paper (100 numbered copies), 10s. 6d.

Hands (T.) Numerical Exercises in Chemistry. Cr. 8vo, **2s. 6d.** and 2s.; Answers separately, 6d.

Hardy (Thomas). See LOW'S STANDARD NOVELS.

Hargreaves (Capt.) **Voyage round** *Great Britain.* Illustrated. Crown 8vo, 5s.

Harland (Marian) Home Kitchen: **a Collection of** *Practical* and Inexpensive Receipts. Crown 8vo, 5s.

Harper's Monthly Magazine. Published Monthly. **160** pages, fully Illustrated. 1s.
    Vol. I. December, 1880, **to** May, 1881.
    ,, II. June to November, 1881.
    ,, III. December, 1881, to May, 1882.
    ,, IV. June to November, 1882.
    ,, V. December, 1882, to May, 1883.
    ,, VI. June to November, 1883.
    ,, VII. December, 1883, to May, 1884.
    ,, VIII. June to November, 1884.
    ,, IX. December, 1884, to May, 1885.
    ,, X. June to November, 1885.
Super-royal 8vo, 8s. 6d. **each.**

"'Harper's Magazine' is so thickly sown with excellent illustrations that to count them would be a work of time; not that it is a picture magazine, for the engravings illustrate the text after the manner seen in some of our choicest *éditions de luxe.*"— *St. James's Gazette.*

"It is so pretty, so big, and so cheap. . . . An extraordinary shillingsworth—160 large octavo pages, with over a score **of** articles, and more than three times as many illustrations."—*Edinburgh Daily Review.*

"An amazing shillingsworth . . . combining choice literature of both nations."—*Nonconformist.*

Harper's Young People. Vol. I., profusely Illustrated with woodcuts and **12** coloured plates. Royal 4to, extra binding, 7s. 6d.; gilt edges, 8s. Published Weekly, in wrapper, **1d.** 12mo. Annual Subscription, **post** free, 6s. **6d.;** Monthly, in wrapper, with coloured plate, 6d.; Annual Subscription, post free, 7s. 6d.

Harrison (Mary) Skilful Cook: a Practical Manual of Modern Experience. Crown 8vo, 5s.

Hatton **(F.)** *North Borneo.* With Biographical Sketch by Jos. **HATTON.** Illustrated from Original Drawings, Map, &c. 8vo, 18s.

Hatton (*Joseph*) *Journalistic London: with Engravings and* Portraits of Distinguished Writers of the **Day.** Fcap. 4to, 12s. 6d.

——— **Three Recruits, and the Girls they left** *behind them.* Small post 8vo, 6s.
    "It hurries us along in unflagging excitement."—*Times.*

Heath **(Francis George)** *Autumnal* **Leaves.** **New** Edition, with Coloured Plates in Facsimile from **Nature.** Crown 8vo, 14s.

——— *Fern Paradise.* New Edition, with Plates and Photos., crown 8vo, 12s. 6d.

Heath (Francis George) Fern World. With Nature-printed Coloured Plates. Crown 8vo, gilt edges, 12s. 6d. Cheap Edition, 6s.

—— *Gilpin's Forest Scenery.* Illustrated, 8vo, 12s. 6d.; New Edition, 7s. 6d.

—— *Our Woodland Trees.* With Coloured Plates and Engravings. Small 8vo, 12s. 6d.

—— *Peasant Life in the West of England.* New Edition, crown 8vo, 10s. 6d.

—— *Sylvan Spring.* With Coloured, &c., Illustrations. 12s. 6d.

—— *Trees and Ferns.* Illustrated, crown 8vo, 3s. 6d.

Heldmann (Bernard) Mutiny on Board the Ship "Leander." Small post 8vo, gilt edges, numerous Illustrations, 5s.

Henty (G. A.) *Winning his* **Spurs.** Illustrations. Cr. 8vo, 5s.

—— *Cornet of Horse: A Story for Boys.* Illust., cr. 8vo, 5s.

—— *Jack Archer: Tale of the Crimea.* Illust., crown 8vo, 5s.

Herrick (Robert) Poetry. Preface by AUSTIN DOBSON. With numerous Illustrations by E. A. ABBEY. 4to, gilt edges, 42s.

Hill (*Staveley, Q.C., M.P.*) *From Home to Home: Two Long* Vacations at the Foot of the Rocky Mountains. With Wood Engravings and Photogravures. 8vo, 21s.

Hitchman, **Public Life of the** *Right Hon. Benjamin Disraeli,* Earl of Beaconsfield. 3rd Edition, with Portrait. Crown 8vo, 3s. 6d.

Holmes (*O. Wendell*) *Poetical Works.* 2 vols., 18mo, exquisitely printed, and chastely bound in limp cloth, gilt tops, 10s. 6d.

Homer. Iliad, **done into** *English Verse.* By A. S. WAY. 5s.

Hudson (W. H.) The Purple Land that England Lost. Travels and Adventures in the Banda-Oriental, South America. 2 vols, crown 8vo, 21s.

Hundred Greatest *Men* (*The*). 8 portfolios, 21s. each, or 4 vols., half-morocco, gilt edges, 10 guineas. New Ed., 1 vol., royal 8vo, 21s.

Hygiene and Public Health. Edited by A. H. BUCK, M.D. Illustrated. 2 vols., royal 8vo, 42s.

Hymnal **Companion of** *Common Prayer.* See BICKERSTETH.

ILLUSTRATED Text-Books of Art-Education. Edited by EDWARD J. POYNTER, R.A. Each Volume contains numerous Illustrations, and is strongly bound for Students, price 5s. Now ready :—

PAINTING.

Classic and Italian. By PERCY R. HEAD.
German, Flemish, and Dutch.
French and Spanish.
English and American.

ARCHITECTURE.

Classic and Early Christian.
Gothic and Renaissance. By T. ROGER SMITH.

SCULPTURE.

Antique: Egyptian and Greek.

Index to the English Catalogue, **Jan.**, 1874, *to* **Dec.**, 1880. Royal 8vo, half-morocco, 18s.

Indian Garden Series. See ROBINSON (**PHIL.**).

Irving (**Henry**) *Impressions of America.* By J. HATTON. 2 vols., 21s.; **New** Edition, 1 vol., **6s.**

Irving (*Washington*). Complete Library Edition of his Works in 27 Vols., Copyright, Unabridged, and with the Author's Latest Revisions, called the "Geoffrey Crayon" Edition, handsomely printed in large square 8vo, on superfine laid paper. Each volume, of about **500 pages,** fully Illustrated. 12s. 6d. per vol. *See also* "Little Britain."

——————————— ("American Men of Letters.") **2s.** 6d.

JAMES (**C.**) *Curiosities of Law and Lawyers.* 8vo, 7s. 6d

Japan. See AUDSLEY.

Jerdon (*Gertrude*) *Key-hole* **Country.** Illustrated. **Crown 8vo,** cloth, 5s.

Johnston (*H. H.*) *River Congo, from its Mouth to Bolobo.* New Edition, 8vo, 21s.

Jones (*Major*) *The Emigrants' Friend.* A Complete Guide to **the** United States. New Edition. 2s. **6d.**

Joyful Lays. Sunday *School Song* **Book.** By LOWRY and DOANE. Boards, 2s.

Julien (*F.*) **English** *Student's French* **Examiner.** 16mo, 2s.

——— *First Lessons* **in** *Conversational French Grammar.* Crown 8vo, 1s.

Julien (F.) French at Home and at School. Book I., Accidence, &c. Square crown 8vo, 2s.

────── *Conversational French Reader.* 16mo, cloth, 2s. 6d.

────── *Petites Leçons de Conversation et de Grammaire.* New Edition, 3s.

────── *Phrases of Daily Use.* Limp cloth, 6d.

KELSEY *(C. B.) Diseases of the Rectum and Anus.* Illustrated. 8vo, 18s.

Kempis (Thomas à) Daily Text-Book. Square 16mo, 2s. 6d.; interleaved as a Birthday Book, 3s. 6d.

Kershaw (S. W.) Protestants from France in their English Home. Crown 8vo, 6s.

Kielland. Skipper Worsé. By the Earl of Ducie. Cr. 8vo, 10s. 6d.

Kingston (W. H. G.) Dick Cheveley. Illustrated, 16mo, gilt edges, 7s. 6d.; plainer binding, plain edges, 5s.

────── *Heir of Kilfinnan.* Uniform, 7s. 6d.; also 5s.

────── *Snow-Shoes and Canoes.* Uniform, 7s. 6d.; also 5s.

────── *Two Supercargoes.* Uniform, 7s. 6d.; also 5s.

────── *With Axe and Rifle.* Uniform, 7s. 6d.; also 5s.

Knight (E. F.) Albania and Montenegro. Illust. 8vo, 12s. 6d.

Knight (E. J.) Cruise of the "Falcon." A Voyage round the World in a 30-Ton Yacht. Illust. New Ed. 2 vols., crown 8vo, 24s.

LANIER *(Sidney) Boy's Froissart.* Illustrated, crown 8vo, gilt edges, 7s. 6d.

────── *Boy's King Arthur.* Uniform, 7s. 6d.

────── *Boy's Mabinogion; Original Welsh Legends of King Arthur.* Uniform, 7s. 6d.

────── *Boy's Percy: Ballads of Love and Adventure, selected from the "Reliques."* Uniform, 7s. 6d.

Lansdell (H.) Through Siberia. 2 vols., 8vo, 30s.; **1 vol.,** 10s. 6d.

―――― *Russia in Central Asia.* Illustrated. **2** vols, **42s.**

Larden (W.) **School Course on Heat.** Second Edition, Illustrated, crown 8vo, **5s.**

Lenormant (F.) Beginnings of History. Crown 8vo, 12s. **6d.**

Leonardo da Vinci's Literary Works. Edited by Dr. JEAN PAUL RICHTER. Containing his Writings on Painting, Sculpture, and Architecture, his Philosophical Maxims, Humorous Writings, and Miscellaneous Notes on Personal Events, on his Contemporaries, on Literature, &c.; published from Manuscripts. 2 vols., imperial 8vo, containing about 200 Drawings in Autotype Reproductions, and numerous other Illustrations. Twelve Guineas.

Library of Religious Poetry. Best Poems of all Ages. Edited by SCHAFF and GILMAN. Royal 8vo, 21s.; re-issue in cheaper binding, 10s. 6d.

Lindsay **(W. S.)** *History of Merchant Shipping.* Over 150 **Illustrations, Maps,** and Charts. In 4 vols., demy 8vo, cloth extra. **Vols. 1 and 2, 11s. each; vols.** 3 and 4, 14s. each. 4 vols., 50s.

Little Britain, The *Spectre Bridegroom,* and *Legend of Sleeepy* Hollow. By WASHINGTON IRVING. An entirely New *Edition de luxe.* Illustrated by 120 very fine Engravings on Wood, by Mr. **J. D.** COOPER. Designed **by** Mr. CHARLES O. MURRAY. Re-issue, **square** crown 8vo, cloth, 6s.

Long **(Mrs.)** *Peace and War in the Transvaal.* **12mo,** 3s. 6d.

Lowell (J. R.) Life of Nathaniel Hawthorn.

Low (Sampson, Jun.) Sanitary Suggestions. Illustrated, **crown** 8vo, 2s. 6d.

Low's Standard Library of Travel **and** *Adventure.* **Crown 8vo,** uniform in cloth extra, 7s. 6d., except where price is given.
  1. **The** Great Lone Land. By Major W. F. BUTLER, C.B.
  2. The Wild North Land. By Major W. F. BUTLER, C.B.
  3. How I found Livingstone. By H. M. STANLEY.
  4. Through the Dark Continent. By H. M. STANLEY. 12s. 6d.
  5. **The** Threshold of the Unknown Region. By C. R. MARKHAM. (4th Edition, with Additional Chapters, 10s. **6d.**)
  6. **Cruise** of the Challenger. By W. J. J. SPRY, R.N.
  7. Burnaby's On Horseback through Asia Minor. 10s. 6d.
  8. Schweinfurth's Heart of Africa. 2 vols., 15s.
  9. Marshall's Through America.
 10. Lansdell's **Through** Siberia. Illustrated and unabridged 10s. 6d.

Low's Standard Novels. Small post 8vo, cloth extra, 6s. each, unless otherwise stated.

A Daughter of Heth. By W. BLACK.
In Silk Attire. By W. BLACK.
Kilmeny. A Novel. By W. BLACK.
Lady Silverdale's Sweetheart. By W. BLACK.
Sunrise. By W. BLACK.
Three Feathers. By WILLIAM BLACK.
Alice **Lorraine**. By R. D. BLACKMORE.
Christowell, a Dartmoor Tale. By R. D. BLACKMORE.
Clara Vaughan. By R. D. BLACKMORE.
Cradock Nowell. By R. D. BLACKMORE.
Cripps the Carrier. By R. D. BLACKMORE.
Erema; or, My Father's Sin. By R. D. BLACKMORE.
Lorna Doone. By R. D. BLACKMORE.
Mary Anerley. By R. D. BLACKMORE.
Tommy Upmore. By R. D. BLACKMORE.
An English Squire. By Miss COLERIDGE.
A Story of the Dragonnades; or, Asylum Christi. By the Rev. E. GILLIAT, M.A.
A Laodicean. By THOMAS HARDY.
Far from the Madding **Crowd**. By THOMAS HARDY.
Pair of Blue Eyes. By THOMAS HARDY.
Return of the Native. By THOMAS HARDY.
The Hand of Ethelberta. By THOMAS HARDY.
The Trumpet **Major**. By THOMAS HARDY.
Two on a Tower. By THOMAS HARDY.
Three Recruits. By JOSEPH HATTON.
A Golden Sorrow. By Mrs. CASHEL HOEY. New Edition.
Out **of Court**. By Mrs. CASHEL HOEY.
Adela Cathcart. By GEORGE MAC DONALD.
Guild Court. **By** GEORGE MAC DONALD.
Mary Marston. By GEORGE MAC DONALD.
Stephen Archer. New Ed. of "Gifts." By GEORGE MAC DONALD.
The Vicar's Daughter. By GEORGE MAC DONALD.
Weighed and Wanting. By GEORGE MAC DONALD.
Diane. By Mrs. MACQUOID.
Elinor Dryden. By Mrs. MACQUOID.
My Lady Greensleeves. By HELEN MATHERS.
Alaric Spenceley. By **Mrs.** J. H. RIDDELL.
Daisies and Buttercups. By Mrs. J. H. RIDDELL.
The Senior Partner. By Mrs. J. H. RIDDELL.
A Struggle for Fame. By Mrs. J. H. RIDDELL.
Jack's Courtship. By W. CLARK RUSSELL.
John Holdsworth. By W. CLARK RUSSELL.
A Sailor's Sweetheart. By W. CLARK RUSSELL.
Sea Queen. **By** W. CLARK RUSSELL.
Watch Below. By W. CLARK RUSSELL.
Wreck of the Grosvenor. By W. CLARK RUSSELL.

Low's Standard Novels—continued.
   The Lady Maud. By W. CLARK RUSSELL.
   Little Loo. By W. CLARK RUSSELL.
   My Wife and I. By Mrs. BEECHER STOWE.
   Poganuc People, their Loves and Lives. By Mrs. B. STOWE.
   Ben Hur: a Tale of the Christ. By LEW. WALLACE.
   Anne. By CONSTANCE FENIMORE WOOLSON.
   For the Major. By CONSTANCE FENIMORE WOOLSON. 5*s.*
   French Heiress in her own Chateau.

Low's Handbook to the Charities of London. Edited and revised to date by C. MACKESON, F.S.S., Editor of "A Guide to the Churches of London and its Suburbs," &c. Yearly, 1*s.* 6*d.*; Paper, 1*s.*

Lyne (Charles) New Guinea. Illustrated, crown 8vo, 10*s.* 6*d.* An Account of the Establishment of the British Protectorate over the Southern Shores of New Guinea.

McCORMICK (*R.*). *Voyages of Discovery in the Arctic and Antarctic Seas in the "Erebus" and "Terror," in Search of* **Sir** John Franklin, &c., with Autobiographical Notice by the Author, who was Medical Officer to each Expedition. With Maps and Lithographic, &c., Illustrations. 2 vols., royal 8vo, 52*s.* 6*d.*

MacDonald (G.) Orts. **Small post 8vo, 6s.**

—— See also " Low's Standard **Novels.**"

Macgregor (John) "Rob Roy" on the Baltic. **3rd Edition,** small post 8vo, 2*s.* 6*d.*; cloth, gilt edges, 3*s.* 6*d.*

—— *A Thousand Miles in the "Rob Roy" Canoe.* 11th Edition, small post 8vo, 2*s.* 6*d.*; cloth, gilt edges, 3*s.* 6*d.*

—— *Voyage Alone in the Yawl "Rob Roy."* New Edition, **with** additions, small post **8vo,** 5*s.*; **3***s.* 6*d.* and **2***s.* 6*d.*

Macquoid (Mrs.). See LOW'S STANDARD NOVELS.

Magazine. See DECORATION, ENGLISH ETCHINGS, HARPER.

Maginn (W.) Miscellanies. **Prose and** *Verse.* With *Memoir.* 2 vols., crown 8vo, 24*s.*

Manitoba. **See** BRYCE.

Manning (E. F.) Delightful Thames. Illustrated. 4to, fancy boards, 5s.

Markham (C. R.) The Threshold of the Unknown Region. Crown 8vo, with Four Maps. 4th Edition. Cloth extra, 10s. 6d.

—— *War between Peru and Chili,* 1879-1881. Third Ed. Crown 8vo, with Maps, 10s. 6d.

—— See also "Foreign Countries."

Marshall (W. G.) Through America. New Ed., cr. 8vo, 7s. 6d.

Martin (F. W.) Float Fishing and Spinning in the Nottingham Style. New Edition. Crown 8vo, 2s. 6d.

Maury (Commander) Physical Geography of the Sea, and its Meteorology. New Edition, with Charts and Diagrams, cr. 8vo, 6s.

Men of Mark: a Gallery of Contemporary Portraits of the most Eminent Men of the Day, specially taken from Life. Complete in Seven Vols., 4to, handsomely bound, cloth, gilt edges, 25s. each.

Mendelssohn Family (The), 1729—1847. From Letters and Journals. Translated. New Edition, 2 vols., 8vo, 30s.

Mendelssohn. See also "Great Musicians."

Merrifield's Nautical Astronomy. Crown 8vo, 7s. 6d.

Millard (H. B.) Bright's Disease of the Kidneys. Illustrated. 8vo, 12s. 6d.

Mitchell (D. G.; Ik. Marvel) Works. Uniform Edition, small 8vo, 5s. each.

Bound together.
Doctor Johns.
Dream Life.
Out-of-Town Places.

Reveries of a Bachelor.
Seven Stories, Basement and Attic.
Wet Days at Edgewood.

Mitford (Mary Russell) Our Village. With 12 full-pape and 157 smaller Cuts. Cr. 4to, cloth, gilt edges, 21s.; cheaper binding, 10s. 6d.

Mollett (J. W.) Illustrated Dictionary of Words used in Art and Archæology. Terms in Architecture, Arms, Bronzes, Christian Art, Colour, Costume, Decoration, Devices, Emblems, Heraldry, Lace, Personal Ornaments, Pottery, Painting, Sculpture, &c. Small 4to, 15s.

Morley (H.) English Literature in the Reign of Victoria. 2000th volume of the Tauchnitz Collection of Authors. 18mo, 2s. 6d.

Morwood (V. S.) Our Gipsies in City, Tent, and Van. 8vo, 18s.

Muller (E.) Noble Words and Noble Deeds. By PHILIPPOTEAUX. Square imperial 16mo, cloth extra, 7s. 6d.; plainer binding, 5s.

Music. See "Great Musicians."

List of Publications. 21

NEW Zealand. See BRADSHAW.

New Zealand Rulers and Statesmen. See GISBORNE.

Newbiggin's **Sketches and** *Tales.* 18mo, 4s.

Nicholls (J. H. Kerry) The King Country: Explorations in New Zealand. Many Illustrations and Map. New Edition, 8vo, 21s.

Nicholson (C.) *Work and* **Workers** *of the British Association.* 12mo, 1s.

Nixon (J.) **Complete** *Story of the Transvaal.* 8vo, 12s. 6d.

Nordhoff (C.) California, for Health, Pleasure, and Residence. New Edition, 8vo, with Maps and Illustrations, 12s. 6d.

Northbrook Gallery. Edited by Lord Ronald Gower. 36 Permanent Photographs. Imperial 4to, 63s.; large paper, 105s.

Nursery Playmates (Prince of). 217 Coloured Pictures for Children by eminent Artists. Folio, in coloured boards, 6s.

O'BRIEN (R. B.) Fifty Years of Concessions to Ireland. With a Portrait of T. Drummond. Vol. I., 16s.; II., 16s.

Orvis (C. F.) Fishing with the Fly. Illustrated. 8vo, 12s. 6d.

Our Little **Ones in** *Heaven.* Edited by the Rev. H. ROBBINS. With Frontispiece after Sir JOSHUA REYNOLDS. New Edition, 5s.

Owen (Douglas) **Marine Insurance** *Notes and* **Clauses.** New Edition, 14s.

PALLISER (Mrs.) A History of Lace. New Edition, with additional cuts and text. 8vo, 21s.

—— *The China Collector's Pocket Companion.* With upwards of 1000 Illustrations of Marks and Monograms. Small 8vo, 5s.

*Pascoe (***C. E.***) London of To-Day.* Illust., crown 8vo, 3s. 6d.

Pharmacopœia of the **United States of** *America.* 8vo, 21s.

Philpot (H. J.) Diabetes Mellitus. Crown 8vo, 5s.

—— *Diet System.* Three Tables, in cases, **1s. each.**

Pinto (Major Serpa) How I Crossed Africa. With 24 full-page and 118 half-page and smaller Illustrations, 13 small Maps, and 1 large one. 2 vols., 8vo, 42*s*.

Plunkett (Major G. F.) Primer of Orthographic Projection. Elementary Practical Solid Geometry clearly explained. With Problems and Exercises. Specially adapted for Science and Art Classes, and for **Students** who have not the aid of **a** Teacher.

Poe (E. A.) The Raven. Illustr. by DORÉ. Imperial folio, 63*s*.

Poems of the Inner Life. Chiefly from Modern Authors. Small 8vo, 5*s*.

Polar Expeditions. See GILDER, MARKHAM, MCCORMICK.

Porter (Noah) Elements of Moral Science. 10*s*. 6*d*.

Powell (W.) Wanderings in a Wild Country; or, Three Years among the Cannibals of New Britain. Illustr., 8vo, 18*s*.; cr. 8vo, 5*s*.

Power (Frank) Letters from Khartoum during the Siege. Fcap. 8vo, boards, 1*s*.

Poynter (Edward J., R.A.). See "Illustrated Text-books."

Publishers' Circular (The), and General Record of British and Foreign Literature. Published on the 1st and 15th of every Month, 3*d*.

*R*EBER *(F.) History of Ancient Art.* 8vo, 18*s*.

Redford (G.) Ancient Sculpture. Crown 8vo, 5*s*.

Richter (Dr. Jean Paul) Italian Art in the National Gallery. 4to. Illustrated. Cloth gilt, 2*l*. 2*s*.; half-morocco, uncut, 2*l*. 12*s*. 6*d*.

———— See also LEONARDO DA VINCI.

Riddell (Mrs. J. H.) See Low's STANDARD NOVELS.

Robin Hood; Merry Adventures of. Written and illustrated by HOWARD PYLE. Imperial 8vo, 15*s*.

Robinson (Phil.) In my Indian Garden. Crown 8vo, limp cloth, 3*s*. 6*d*.

Robinson (Phil.) Indian Garden Series. 1s **6d.**; boards, 1s. each.
I. Chasing a Fortune, &c. : Stories. II. Tigers **at Large.**

—— —— *Noah's Ark. A Contribution to the Study of Unnatural History.* Small post 8vo, 12s. 6d.

—— —— *Sinners and Saints: a Tour across the United States of America, and Round them.* Crown 8vo, 10s. 6d.

—— —— *Under the Punkah.* Crown 8vo, limp cloth, 5s.

Rockstro (W. S.) History of Music.

Rodrigues (J. C.) The Panama Canal. Crown 8vo, cloth extra, 5s.

"A series of remarkable articles . . . a mine of valuable data for editors and diplomatists."—*New York Nation.*

Roland; the Story of. Crown 8vo, illustrated, 6s.

Rose (J.) Complete Practical Machinist. New Ed., 12mo, 12s. 6d.

—— —— *Mechanical Drawing.* Illustrated, small 4to, 16s.

Rose Library (The). Popular Literature of all Countries. Each volume, 1s.; **cloth, 2s.** 6d. Many of the Volumes are Illustrated—

Little **Women.** By LOUISA M. ALCOTT.

Little Women Wedded. Forming a Sequel to "Little Women."

Little Women and Little Women Wedded. 1 vol., cloth gilt, 3s. 6d.

Little **Men.** By L. M. ALCOTT. **2s.**; cloth gilt, **3s. 6d.**

An Old-Fashioned Girl. By LOUISA M. ALCOTT. **2s.**; cloth, 3s. 6d.

Work. A Story of Experience. By L. M. ALCOTT. **3s.** 6d.; **2 vols.** 1s. each.

Stowe (Mrs. H. B.) The Pearl of Orr's Island.

—— —— The Minister's Wooing.

—— —— We and our Neighbours. 2s.; cloth **gilt, 6s.**

—— —— My Wife and I. 2s.; cloth gilt, 6s.

Hans Brinker; **or, the** Silver Skates. By Mrs. DODGE.

My Study Windows. By J. R. LOWELL.

The Guardian Angel. By OLIVER WENDELL HOLMES.

My Summer in a Garden. By **C.** D. WARNER.

Dred. By Mrs. BEECHER STOWE. 2s.; cloth gilt, **3s. 6d.**

Farm Ballads. By WILL CARLETON.

Farm Festivals. By WILL CARLETON.

Rose Library (The)—continued.

 Farm Legends. By WILL CARLETON.
 The Clients of Dr. Bernagius. 3*s.* 6*d.* ; 2 parts, 1*s.* each.
 The Undiscovered Country. By W. D. HOWELLS. 3*s.* 6*d.* and 1*s.*
 Baby Rue. By C. M. CLAY. 3*s.* 6*d.* and 1*s.*
 The Rose in Bloom. By L. M. ALCOTT. 2*s.* ; cloth gilt, 3*s.* 6*d.*
 Eight Cousins. By L. M. ALCOTT. 2*s.* ; cloth gilt, 3*s.* 6*d.*
 Under the Lilacs. By L. M. ALCOTT. 2*s.* ; also 3*s.* 6*d.*
 Silver Pitchers. By LOUISA M. ALCOTT. 3*s.* 6*d.* and 1*s.*
 Jimmy's Cruise in the "Pinafore," and other Tales. By LOUISA M. ALCOTT. 2*s.*; cloth gilt, 3*s.* 6*d.*
 Jack and Jill. By LOUISA M. ALCOTT. 5*s.*; 2*s.*
 Hitherto. By the Author of the "Gayworthys." 2 vols., 1*s.* each; 1 vol., cloth gilt, 3*s.* 6*d.*
 Friends: a Duet. By E. STUART PHELPS. 3*s.* 6*d.*
 A Gentleman of Leisure. A Novel. By EDGAR FAWCETT. 3*s.* 6*d.* ; 1*s.*
 The Story of Helen Troy. 3*s.* 6*d.*; also 1*s.*

Ross (Mars; and Stonehewer Cooper) Highlands of Cantabria; or, Three Days from England. Illustrations and Map, 8vo, 21*s.*

Round the Yule Log: Norwegian Folk and Fairy Tales. Translated from the Norwegian of P. CHR. ASBJÖRNSEN. With 100 Illustrations after drawings by Norwegian Artists, and an Introduction by E. W Gosse. Impl. 16mo, cloth extra, gilt edges, 7*s.* 6*d.* and 5*s.*

Rousselet (Louis) Son of the Constable of France. Small post 8vo, numerous Illustrations, 5*s.*

—— *King of the Tigers : a Story of Central India.* Illustrated. Small post 8vo, gilt, 6*s.*; plainer, 5*s.*

—— *Drummer Boy.* Illustrated. Small post 8vo, 5*s.*

Rowbotham (F.) Trip to Prairie Land. The Shady Side of Emigration. 5*s.*

Russell (W. Clark) English Channel Ports and the Estate of the East and West India Dock Company. Crown 8vo, 1*s.*

—— *Jack's Courtship.* 3 vols., 31*s.* 6*d.* ; 1 vol., 6*s.*

Russell (W. Clark) The Lady Maud. **3 vols.,** 31*s.* 6*d.*; **1 vol.,** 6*s.*

—— —— *Little Loo.* **New** Edition, small post 8vo, 6*s.*

—— —— *My Watch Below; or, Yarns Spun when off Duty.* Small post 8vo, 6*s.*

—— —— *Sailor's Language.* Illustrated. Crown 8vo, 3*s.* 6*d.*

—— —— *Sea Queen.* 3 vols., 31*s.* 6*d.*; 1 vol., 6*s.*

—— —— *Strange Voyage.* Nautical Novel. 3 vols., crown 8vo, 31*s.* 6*d.*

—— —— *Wreck of the Grosvenor.* **4to**, sewed, **6*d.***

—— —— See also Low's STANDARD NOVELS.

*S*AINTS *and their Symbols: A Companion in the Churches* and Picture Galleries of Europe. Illustrated. Royal 16mo, 3*s.* 6*d.*

Salisbury (Lord) Life and Speeches. By F. S. Pulling, M.A. With Photogravure Portrait of Lord Salisbury. 2 vols., crown 8vo, 21*s.*

Saunders (A.) **Our Domestic Birds:** *Poultry in England and New Zealand.* **Crown 8vo, 6*s.***

Scherr (Prof. J.) History of English *Literature.* Cr. 8vo, 8*s.* 6*d.*

Schley. Rescue of Greely. Maps and Illustrations, 8vo, 12*s.* 6*d.*

Schuyler (Eugène). **The** *Life of Peter the Great.* By EUGÈNE SCHUYLER, Author of "Turkestan." 2 vols., 8vo, 32*s.*

Schweinfurth (Georg) Heart of Africa. Three Years' Travels and Adventures in the Unexplored Regions of Central Africa, from 1868 to 1871. Illustrations and large Map. 2 vols., crown 8vo, **15*s.***

Scott (Leader) Renaissance of Art in Italy. 4to, 31*s.* 6*d.*

Sea, *River, and Creek.* By GARBOARD STREYKE. *The Eastern Coast.* 12mo, 1*s.*

Senior (W.) Waterside **Sketches.** Imp. 32mo, 1*s.*6*d.*, boards, 1*s.*

Shadbolt and Mackinnon's South African Campaign, 1879. Containing a portrait and biography of every officer who lost his life. 4to, handsomely bound, 2*l.* 10*s.*

Shadbolt (S. H.) Afghan Campaigns of 1878—1880. By SYDNEY SHADBOLT. 2 vols., royal quarto, cloth extra, 3*l.*

Shakespeare. Edited by R. GRANT WHITE. 3 vols., crown 8vo, gilt top, 36*s.*; *édition de luxe*, 6 vols., 8vo, cloth extra, 63*s.*

Shakespeare. See also WHITE (R. GRANT).

"*Shooting Niagara;*" *or, The Last Days of Caucusia.* By the Author of "The New Democracy." Small post 8vo, boards, 1*s.*

Sidney (Sir Philip) Arcadia. New Edition, 6*s.*

Siegfried : The Story of. Illustrated, crown 8vo, cloth, 6*s.*

Sinclair (Mrs.) Indigenous Flowers of the Hawaiian Islands. 44 Plates in Colour. Imp. folio, extra binding, gilt edges, 31*s.* 6*d.*

Sir Roger de Coverley. Re-imprinted from the "Spectator." With 125 Woodcuts and special steel Frontispiece. Small fcap. 4to, 6*s.*

Smith (G.) Assyrian Explorations and Discoveries. Illustrated by Photographs and Woodcuts. New Edition, demy 8vo, 18*s.*

———— *The Chaldean Account of Genesis.* With many Illustrations. 16*s.* New Edition, revised and re-written by PROFESSOR SAYCE, Queen's College, Oxford. 8vo, 18*s.*

Smith (J. Moyr) Ancient Greek Female Costume. 112 full-page Plates and other Illustrations. Crown 8vo, 7*s.* 6*d.*

———— *Hades of Ardenne: a Visit to the Caves of Han.* Crown 8vo, Illustrated, 5*s.*

———— *Legendary Studies, and other Sketches for Decorative* Figure Panels. 7*s.* 6*d.*

———— *Wooing of Æthra.* Illustrated. 32mo, 1*s.*

Smith (Sydney) Life and Times. By STUART J. REID. Illustrated. 8vo, 21*s.*

Smith (T. Roger) Architecture, Gothic and Renaissance. Illustrated, crown 8vo, 5*s.*

———————————— *Classic and Early Christian.* Illustrated. Crown 8vo, 5*s.*

Smith (W. R.) Laws concerning Public Health. 8vo, 31*s.* 6*d.*

Somerset (*Lady H.*) *Our Village Life.* Words and Illustrations. Thirty Coloured Plates, royal 4to, fancy covers, 5s.

Spanish and French **Artists.** By **Gerard Smith.** (Poynter's Art Text-books.) 5s.

Spiers' French Dictionary. **29th** Edition, remodelled. **2** vols., 8vo, 18s.; half bound, 21s.

Spry (*W. J. J., R.N.*) *Cruise of H.M.S. "Challenger."* With many Illustrations. 6th Edition, 8vo, cloth, 18s. Cheap Edition, crown 8vo, 7s. 6d.

Spyri (*Joh.*) *Heidi's Early Experiences: a Story for Children* and those who love Children. Illustrated, small post 8vo, 4s. 6d.

———— *Heidi's Further Experiences.* Illust., sm. post 8vo, 4s. **6d.**

Stanley (*H. M.*) *Congo, and Founding its Free State.* Illustrated, 2 vols., 8vo, 42s.

———— *How I Found Livingstone.* 8vo, 10s. 6d.; cr. 8vo, 7s. 6d.

———— *Through the Dark Continent.* Crown 8vo, 12s. 6d.

Stenhouse (*Mrs.*) *An Englishwoman in Utah.* Crown 8vo, 2s. 6d.

Stevens (*E. W.*) *Fly-Fishing in Maine Lakes.* 8s. 6d.

Stockton (*Frank R.*) *The* **Story of Viteau.** With 16 page Illustrations. **Crown** 8vo, 5s.

Stoker (*Bram*) *Under the Sunset.* Crown 8vo, **6s.**

Stowe (*Mrs. Beecher*) *Dred.* Cloth, gilt edges, 3s. 6d.; boards, 2s.

———— *Little Foxes.* Cheap Ed., 1s.; Library Edition, 4s. 6d.

———— *My Wife and I.* Small post 8vo, 6s.

———— *Old Town Folk.* 6s.; Cheap Edition, 3s.

———— **Old** *Town Fireside Stories.* Cloth extra, 3s. 6d.

———— *We and our* **Neighbours.** Small post 8vo, 6s.

———— *Poganuc* **People: their Loves and Lives.** Crown 8vo, 6s.

———— **Chimney Corner. 1s.**; cloth, 1s. 6d.

———— See also Rose Library.

Sullivan (A. M.) Nutshell History of Ireland. Paper boards, 6d.
Sutton (A. K.) A B C Digest of the Bankruptcy Law. 8vo, 3s. and 2s. 6d.

TAINE (H. A.) "Les Origines de la France Contemporaine." Translated by JOHN DURAND.
    I. The Ancient Regime. Demy 8vo, cloth, 16s.
    II. The French Revolution. Vol. 1. do.
    III. Do. do. Vol. 2. do.
    IV. Do. do. Vol. 3. do.

Talbot (Hon. E.) A Letter on Emigration. 1s.
Tauchnitz's English Editions of German Authors. Each volume, cloth flexible, 2s.; or sewed, 1s. 6d. (Catalogues post free.)
Tauchnitz (B.) German and English Dictionary. 2s.; paper, 1s. 6d.; roan, 2s. 6d.
—— *French and English Dictionary.* 2s.; paper, 1s. 6d.; roan, 2s. 6d.
—— *Italian and English Dictionary.* 2s.; paper, 1s. 6d.; roan, 2s. 6d.
—— *Spanish and English.* 2s.; paper, 1s. 6d.; roan, 2s. 6d.
Taylor (W. M.) Paul the Missionary. Crown 8vo, 7s. 6d.
Thausing (Prof.) Malt and the Fabrication of Beer. 8vo, 45s.
Theakston (M.) British Angling Flies. Illustrated. Cr. 8vo, 5s.
Thomson (W.) Algebra for Colleges and Schools. With numerous Examples. 8vo, 5s., Key, 1s. 6d.
Thomson (Jos.) Through Masai Land. Illustrations and Maps. 21s.
Thoreau. American Men of Letters. Crown 8vo, 2s. 6d.
Tolhausen (Alexandre) Grand Supplément du Dictionnaire Technologique. 3s. 6d.
Tristram (Rev. Canon) Pathways of Palestine: A Descriptive Tour through the Holy Land. First Series. Illustrated by 44 Permanent Photographs. 2 vols., folio, cloth extra, gilt edges, 31s. 6d. each.

Trollope (Anthony) Thompson Hall. 1s.

Tromholt (S.) Under the **Rays of the Aurora** Borealis. By C. SIEWERS. Photographs and Portraits. 2 vols., 8vo, 30s.

Tunis. See REID.

Turner (Edward) Studies in Russian Literature. Cr. 8vo, 8s. 6d.

UNION Jack (The). Every Boy's Paper. Edited by G. A. HENTY. Profusely Illustrated with Coloured and other Plates. Vol. I., 6s. Vols. II., III., IV., 7s. 6d. each.

VASILI (Count) Berlin Society. Translated. Cown 8vo, 6s.

——— *World of* London *(La Société de Londres).* Translated. Crown 8vo, 6s.

Velazquez and Murillo. By C. B. CURTIS. With Original Etchings. Royal 8vo, 31s. 6d.; large paper, 63s.

Victoria (Queen) Life of. By GRACE GREENWOOD. With numerous Illustrations. **Small post 8vo, 6s.**

Vincent (Mrs. *Howard) Forty Thousand Miles over Land and Water.* With Illustrations engraved under the direction of Mr. H. BLACKBURN. 2 vols, crown 8vo, 21s.

Viollet-le-Duc *(E.)* **Lectures** *on Architecture.* Translated by BENJAMIN BUCKNALL, Architect. With 33 **Steel** Plates and 200 Wood Engravings. Super-royal 8vo, leather back, **gilt** top, 2 vols., 3l. 3s.

Vivian *(A. P.) Wanderings in the Western Land.* 3rd Ed., 10s. 6d.

BOOKS BY JULES VERNE.

WORKS.	LARGE CROWN 8vo.	Containing 350 to 600 pp. and from 50 to 100 full-page illustrations.		Containing the whole of the text with some illustrations.	
		In very handsome cloth binding, gilt edges.	In plainer binding, plain edges.	In cloth binding, gilt edges, smaller type.	Coloured boards.
		s. d.	s. d.	s. d.	
20,000 Leagues under the Sea. Parts I. and II.		10 6	5 0	3 6	2 vols., 1s. each.
Hector Servadac		10 6	5 0	3 6	2 vols., 1s. each.
The Fur Country		10 6	5 0	3 6	2 vols., 1s. each.
The Earth to the Moon and a Trip round it		10 6	5 0	{ 2 vols., 2s. ea. }	2 vols., 1s. each.
Michael Strogoff		10 6	5 0	3 6	2 vols., 1s. each.
Dick Sands, the Boy Captain		10 6	5 0	3 6	2 vols., 1s. each.
Five Weeks in a Balloon		7 6	3 6	2 0	1s. 0d.
Adventures of Three Englishmen and Three Russians		7 6	3 6	2 0	1 0
Round the World in Eighty Days		7 6	3 6	2 0	1 0
A Floating City		7 6	3 6	2 0	1 0
The Blockade Runners		7 6	3 6	2 0	1 0
Dr. Ox's Experiment		—	—	2 0	1 0
A Winter amid the Ice		—	—	2 0	1 0
Survivors of the "Chancellor"		7 6	3 6	2 0	2 vols., 1s. each.
Martin Paz		7 6	3 6	2 0	1s. 0d.
The Mysterious Island, 3 vols.:—		22 6	10 6	6 0	3 0
I. Dropped from the Clouds		7 6	3 6	2 0	1 0
II. Abandoned		7 6	3 6	2 0	1 0
III. Secret of the Island		7 6	3 6	2 0	1 0
The Child of the Cavern		7 6	3 6	2 0	1 0
The Begum's Fortune		7 6	3 6	2 0	1 0
The Tribulations of a Chinaman		7 6	3 6	2 0	1 0
The Steam House, 2 vols.:—					
I. Demon of Cawnpore		7 6	3 6	2 0	1 0
II. Tigers and Traitors		7 6	3 6	2 0	1 0
The Giant Raft, 2 vols.:—					
I. 800 Leagues on the Amazon		7 6	3 6	2 0	1 0
II. The Cryptogram		7 6	3 6	2 0	1 0
The Green Ray		6 0	5 0	—	1 0
Godfrey Morgan		7 6	3 6	2 0	1 0
Kéraban the Inflexible:—					
I. Captain of the "Guidara"		7 6			
II. Scarpante the Spy		7 6			
The Archipelago on Fire		7 6			
The Vanished Diamond		7 6			

CELEBRATED TRAVELS AND TRAVELLERS. 3 vols. 8vo, 600 pp., 100 full-page illustrations, 12s. 6d.; gilt edges, 14s. each:—(1) THE EXPLORATION OF THE WORLD. (2) THE GREAT NAVIGATORS OF THE EIGHTEENTH CENTURY. (3) THE GREAT EXPLORERS OF THE NINETEENTH CENTURY.

WAHL (W. H.) *Galvanoplastic Manipulation for the Electro-Plater.* 8vo, 35s.

Wallace (L.) *Ben Hur: A Tale of the Christ.* Crown 8vo, **6s.**

Waller (Rev. C. H.) *The Names on the Gates of Pearl*, and other Studies. New Edition. Crown 8vo, cloth extra, 3s. 6d.

———— *A Grammar and Analytical Vocabulary of the Words in the Greek Testament.* Compiled from Brüder's Concordance. For the use of Divinity Students and Greek Testament Classes. Part I. Grammar. Small post 8vo, cloth, 2s. 6d. Part II. Vocabulary, 2s. 6d.

———— *Adoption and the Covenant.* Some Thoughts on Confirmation. Super-royal 16mo, cloth limp, **2s. 6d.**

———— *Silver Sockets; and other Shadows of Redemption.* Sermons at Christ Church, Hampstead. Small post 8vo, 6s.

Walton (Iz.) *Wallet Book*, CIƆIƆLXXXV. 21s.; l. p. **42s.**

Walton (T. **H.**) *Coal Mining.* With Illustrations. 4to, 25s.

Warder (**G. W.**) *Utopian Dreams and Lotus Leaves.* Crown 8vo, 6s.

Warner (C. D.) *My **Summer in a Garden.*** Boards, 1s.; leatherette, 1s. 6d.; cloth, **2s.**

Warren (W. F.) *Paradise Found; the North Pole **the** Cradle of the Human Race.* Illustrated. Crown 8vo, 12s. 6d.

Washington Irving's *Little Britain.* Square crown 8vo, 6s.

Watson (P. B.) *Marcus Aurelius Antoninus.* Portr. 8vo, 15s.

Webster. (American Men of Letters.) 18mo, 2s. 6d.

Weir (Harrison) *Animal Stories, Old and New, told in Pictures* **and** Prose. Coloured, &c., Illustrations. 56 pp., 4to, 5s.

Wells (H. P.) *Fly Rods and Fly Tackle.* Illustrated. **10s. 6d.**

Wheatley (H. B.) and Delamotte (P. H.) *Art Work* **in** *Porcelain.* Large 8vo, 2s. 6d.

———— *Art Work in Gold and Silver. Modern.* Large **8vo, 2s. 6d.**

———— *Handbook of Decorative Art.* **10s. 6d.**

Whisperings. Poems. **Small** post 8vo, **cloth extra,** gilt edges, 3s. 6d.

White (R. **Grant**) ***England*** *Without and* ***Within.*** Crown 8vo, 10s. 6d.

———— *Every-day English.* Crown 8vo, 10s. 6d.

———— *Studies in Shakespeare.* Crown 8vo, 10s. 6d.

White (R. Grant) Fate of Mansfield Humphreys, the Episode of Mr. Washington Adams in England, an Apology, &c. Crown 8vo, 6s.

—— *Words and their uses.* New Edit., crown 8vo, 10s. 6d.

Whittier (J. G.) The King's Missive, and later Poems. 18mo, choice parchment cover, 3s. 6d.

—— *The Whittier Birthday Book.* Extracts from the Author's writings, with Portrait and Illustrations. Uniform with the "Emerson Birthday Book." Square 16mo, very choice binding, 3s. 6d.

—— *Life of.* By R. A. UNDERWOOD. Cr. 8vo, cloth, 10s. 6d.

Williams (C. F.) Tariff Laws of the United States. 8vo, 10s. 6d.

Williams (H. W.) Diseases of the Eye. 8vo, 21s.

Wills, A Few Hints on Proving, without Professional Assistance. By a PROBATE COURT OFFICIAL. 8th Edition, revised, with Forms of Wills, Residuary Accounts, &c. Fcap. 8vo, cloth limp, 1s.

Wimbledon (Viscount) Life and Times, 1628-38. By C. DALTON. 2 vols., 8vo, 30s.

Witthaus (R. A.) Medical Student's Chemistry. 8vo, 16s.

Woodbury, History of Wood Engraving. Illustrated. 8vo, 18s.

*Woolsey (C. D., LL.D.) Introduction to the Study of Inter-*national Law. 5th Edition, demy 8vo, 18s.

Woolson (Constance F.) See "Low's Standard Novels."

Wright (H.) Friendship of God. Portrait, &c. Crown 8vo, 6s.

Written to Order; the Journeyings of an Irresponsible Egotist. Crown 8vo, 6s.

*Y*RIARTE *(Charles) Florence: its History.* Translated by C. B. PITMAN. Illustrated with 500 Engravings. Large imperial 4to, extra binding, gilt edges, 63s.; or 12 Parts, 5s. each.

History; the Medici; the Humanists; letters; arts; the Renaissance; illustrious Florentines; Etruscan art; monuments; sculpture; painting.

London:
SAMPSON LOW, MARSTON, SEARLE, & RIVINGTON,
CROWN BUILDINGS, 188, FLEET STREET, E.C.

www.ingramcontent.com/pod-product-compliance
Lightning Source LLC
Chambersburg PA
CBHW022059230426
43672CB00008B/1218